FOURIER ANALYSIS ON GROUPS AND PARTIAL WAVE ANALYSIS

MATHEMATICS LECTURE NOTE SERIES

J. Frank Adams	LECTURES ON LIE GROUPS
E. Artin and J. Tate	CLASS FIELD THEORY
Michael Atiyah	K-THEORY
Jacob Barshay	TOPICS IN RING THEORY
Hyman Bass	ALGEBRAIC K-THEORY
Melvyn S. Berger Marion S. Berger	PERSPECTIVES IN NONLINEARITY
Armand Borel	LINEAR ALGEBRA GROUPS
Raoul Bott	LECTURES ON K (X)
Andrew Browder	INTRODUCTION TO FUNCTION ALGEBRAS

Gustave Choquet	LECTURES ON ANALYSIS
	  I. INTEGRATION AND TOPOLOGICAL VECTOR SPACES
	  II. REPRESENTATION THEORY
	  III. INFINITE DIMENSIONAL MEASURES AND PROBLEM SOLUTIONS

Paul J. Cohen	SET THEORY AND THE CONTINUUM HYPOTHESIS
Eldon Dyer	COHOMOLOGY THEORIES
Robert Ellis	LECTURES ON TOPOLOGICAL DYNAMICS
Walter Feit	CHARACTERS OF FINITE GROUPS
John Fogarty	INVARIANT THEORY
William Fulton	ALGEBRAIC CURVES
Marvin J. Greenberg	LECTURES ON ALGEBRAIC TOPOLOGY
Marvin J. Greenberg	LECTURES ON FORMS IN MANY VARIABLES
Robin Hartshorne	FOUNDATIONS OF PROJECTIVE GEOMETRY
Robert Hermann	FOURIER ANALYSIS ON GROUPS AND PARTIAL WAVE ANALYSIS

J. F. P. Hudson	PIECEWISE LINEAR TOPOLOGY
Irving Kaplansky	RINGS OF OPERATORS
K. Kapp and H. Schneider	COMPLETELY O-SIMPLE SEMIGROUPS
Joseph B. Keller	BIFURCATION THEORY AND
Stuart Antman	NONLINEAR EIGENVALUE PROBLEMS
Serge Lang	ALGEBRAIC FUNCTIONS
Serge Lang	RAPPORT SUR LA COHOMOLOGIE DES GROUPES
Ottmar Loos	SYMMETRIC SPACES I. GENERAL THEORY II. COMPACT SPACES AND CLASSIFICATIONS
I. G. Macdonald	ALGEBRAIC GEOMETRY: INTRODUCTION TO SCHEMES
George W. Mackey	INDUCED REPRESENTATIONS OF GROUPS AND QUANTUM MECHANICS
Andrew Ogg	MODULAR FORMS AND DIRICHLET SERIES
Richard Palais	FOUNDATIONS OF GLOBAL NON-LINEAR ANALYSIS
William Parry	ENTROPY AND GENERATORS IN ERGODIC THEORY
D. S. Passman	PERMUTATION GROUPS
Walter Rudin	FUNCTION THEORY IN POLYDISCS
Jean-Pierre Serre	ABELIAN l-ADIC REPRESENTATIONS AND ELLIPTIC CURVES
Jean-Pierre Serre	ALGEBRES DE LIE SEMI-SIMPLE COMPLEXES
Jean-Pierre Serre	LIE ALGEBRAS AND LIE GROUPS
Shlomo Sternberg	CELESTIAL MECHANICS PART I
Shlomo Sternberg	CELESTIAL MECHANICS PART II
Moss E. Sweedler	HOPF ALGEBRAS

A Note from the Publisher

This volume was printed directly from a typescript prepared by the author, who takes full responsibility for its content and appearance. The Publisher has not performed his usual functions of reviewing, editing, typesetting, and proofreading the material prior to publication.

The Publisher fully endorses this informal and quick method of publishing lecture notes at a moderate price, and he wishes to thank the author for preparing the material for publication.

FOURIER ANALYSIS ON GROUPS AND PARTIAL WAVE ANALYSIS

ROBERT HERMANN

Institute of Advanced Study

Princeton, New Jersey

W. A. BENJAMIN, INC.

New York   1969

FOURIER ANALYSIS ON GROUP AND PARTIAL
WAVE ANALYSIS

Copyright ©1969 by W. A. Benjamin, Inc.
All rights reserved

Standard Book Numbers   8053-3940-x (Clothbound)
                         8053-3941-8 (Paperback)
Library of Congress Catalog Card Number: 70-104383
Manufactured in the United States of America
12345K32109

*The manuscript was put into production on October 1, 1969;
this volume was published on December 15, 1969*

W. A. BENJAMIN, INC.
New York, New York 10016

FOURIER ANALYSIS ON GROUPS
AND PARTIAL WAVE ANALYSIS

PREFACE

Although it is not generally known among mathematicians, contemporary elementary particle physics presents many fascinating mathematical problems in the area of Lie group theory and differential geometry. In these notes we present material concerning Lie group representation theory that is most closely linked to the study of those general properties of the scattering operator that follow from its invariance under the symmetries of the underlying physical problem. (This is, of course, the reference to "partial wave analysis" in the title.) Additional topics in differential geometry and Lie algebra theory more closely related to quantum mechanics and quantum field theory proper are presented in a companion volume, "Lie Algebras and Quantum Mechanics."

This material may be considered as a sequel to my book, "Lie Groups for Physicists." Although in these notes one is involved with details that are in much less definite form than that in "Lie Groups for Physicists," my aim is again to present and develop mathematical material that is relevant for physics in a style that combines certain features of that used by the mathematician and physicist. I realize that I thereby risk offending individuals in both groups. On the one hand, in these notes I do not follow the standards of absolute precision and rigor that is now customary in mathematical literature. Unfortunately, this insistence on a rigid style has been a substantial contributing factor to the separation of mathematics and physics, and to the de-emphasis on intuition and creative imagination in contemporary mathematics. On the other hand, I do not subscribe either to the physicists' "traditional" viewpoint that mathematics is a necessary evil in understanding Nature, to be dealt with on as lowbrow a level as possible. Indeed, I admit that it gives me a certain esthetic pleasure to see sophisticated mathematical concepts enter very naturally into physical problems, and I regard the development of this connection to be one of the major tasks of contemporary mathematical physics. There has been a remarkable flowering of mathematics in the last twenty years -- a period comparable to the 1905-1930 in theoretical physics -- and the next major task is to reintegrate this with the mainstream of science.

I would like to thank the typists, Carol Tung and Alta Zapf.

CONTENTS

PREFACE v

CHAPTER I

Meromorphic Decompositions and Analytic Continuation

    1. Introduction 1
    2. Hilbert and Dirac Spaces 2
    3. Reduction from SO(2,1) to SO(1,1) 7
    4. Asymptotic Expansions Obtained from Spectral Resolutions 12
    5. A Possible Generalization for Subgroup Decompositions 18
    6. Reduction from SL(2,C) to SL(2,R) 21
    7. A General Formulation of the Reduction Procedure 29
    8. Limits and Contractions of Lie Groups and the t=0 Symmetry 35
    9. Analytic Continuation of the Fourier Expansion of Functions on SO(3,R) and E(2) 39
    Bibliography 48

CHAPTER II

The Fourier Transform on Lie Groups

    Introduction 51
    1. Fourier Transform on Groups 52
    2. The Generalized-Function Approach to the Fourier Transform 59
    3. Integrals on Homogeneous Spaces 64
    4. The Fourier Transform as an Integral Operator 70
    5. Characters of Multiplier Representations 73

6. Relations Between the Fourier
   Expansions of Functions on
   Different Real Forms of the Same
   Complex Group   76
7. Properties of the Cauchy Kernel   80
Bibliography   86

CHAPTER III

Cauchy Integrals on Lie Groups and Matrix
Element Functions of the Second Kind

1. Introduction   87
2. General Remarks Concerning Cauchy
   Kernels for Complex Lie Groups   91
3. Hua's Construction of Cauchy
   Kernels   94
4. Expansions of Cauchy Kernels
   over Compact Groups and Wigner
   d-Functions of the Second Kind   98
5. Group-Theoretic Meaning of the
   Integral Representations for the
   Legendre Functions of the
   Second Kind   101
6. Matrix Element Functions of the
   Second Kind for General Groups   112
Bibliography   115

CHAPTER IV

Deformation of the Fourier Integral on Groups
From Compact to Non-Compact Groups

1. Introduction   117
2. Limiting Relations for Subgroups
   of $SO(2,1)$ 118
3. Direct Passage from a Sum to
   an Integral   125
4. Deformation of the Fourier
   Expansion from $SO(3,R)$ to $E(2)$ 128
Bibliography   138

CHAPTER V

Partial Wave Analysis of the Scattering Amplitude

1.	Some General Principles of Group Representations Theory	140
2.	Irreducible Representations of Semidirect Products and the Poincaré Group	147
3.	Partial Wave Analysis of the Scattering Amplitude	156
	Bibliography	177

CHAPTER VI

Partial Wave Analysis as a Problem in Group Representation Theory

1.	Introduction	179
2.	A General Algebraic Viewpoint in Functional Analysis Suggested by S-Matrix Theory	180
3.	The Irreducible Projection Operators Defined by Integration Over the Group	190
4.	Crossing Symmetry in a Group-Theoretic Framework	197
	Bibliography	206

CHAPTER VII

Remarks on the Use of Transformation Groups in Quantum Field Theory

1.	Connections Between Vector Bundles and Quantum Fields	209
2.	Yang-Mills Fields and the Differential Geometry of Vector Bundles	221
3.	Gauge Transformation of Linear Connections; a Cohomology Interpretation	225

4.	The Curvature of a Vector-Bundle Connection	229
5.	Gauge Invariant Couplings Between Cross-Sections and Connections	231
Bibliography		236

CHAPTER VIII

Generalized Functions on Manifolds

1.	Introduction	239
2.	Dirac Spaces	240
3.	Dirac Spaces Manifolds and Their Behavior Under Mappings	246
4.	Generalized Functions Associated with Mappings of Pseudoriemannian Manifolds	253
5.	Generalized Functions Defined by Hypersurfaces	261
6.	Generalized Functions Generated by Flows	271
7.	Remarks on the Behavior of Generalized Functions Under General Mappings	280
8.	Generalized Functions on Fibre Spaces	290
Bibliography		301

CHAPTER I
MEROMORPHIC DECOMPOSITIONS AND ANALYTIC CONTINUATION

1. INTRODUCTION

Work in recent years by Boyce, Domokos and Suranyi, Freedman and Wang, Hadjiannou, Joos, Sciarrino, Sertorio and Toller has developed a point of view towards the analysis of the S-matrix that emphasizes the connections with group-representation theory and the harmonic analysis of functions on groups. The goal of these notes is to survey and develop some of the mathematical tools and ideas that seem to be relevant to this analysis, particularly emphasizing and influenced by the work of Toller and his collaborators. In this first chapter we cover primarily two topics: (a) the mathematical background for the calculation done by Sciarrino and Toller [11] whereby a single "Lorentz pole" in the forward scattering case is resolved into a sequence of integer-spaced Regge poles; (b) a mathematical problem (involving a transition from the Plancherel formula from a compact to a noncompact group) which seems to mirror the physical problem involved in unequal-mass elastic scattering in the neighborhood of the $t = 0$ point.

I would like to thank G. Domokos and M. Toller, with whom I have had valuable discussions about some of this material.

2. HILBERT AND DIRAC SPACES

Let H be an (incomplete) Hilbert space; denote elements by Ψ, the Hermitian inner product by

$$\langle \Psi | \Psi' \rangle.$$

Let $\underset{\sim}{D}$ be the space of <u>all</u> linear functionals defined on H. It is called the *Dirac space* associated with H. We will adopt a notation that is close to that used by Dirac in his book on quantum mechanics.

Denote an element of $\underset{\sim}{D}$ by α. Define $\langle\alpha|\Psi\rangle$ as the value $\alpha(\Psi)$ of the functional α on Ψ. Let

$$\langle\Psi|\alpha\rangle = \langle\alpha|\Psi\rangle^* \quad (* = \text{complex conjugate}).$$

If $c \in C$ (= complex numbers), define $c\alpha$ as the linear functional $\Psi \to c^*\alpha(\Psi)$. This makes $\underset{\sim}{D}$ into a complex vector space. Further, the assignment of the linear functional

$$\Psi' \to \langle\Psi|\Psi'\rangle$$

to each $\Psi \in H$ defines an embedding of $H \to \underset{\sim}{D}$ that (with this convention) is complex linear.

Let A be an operator $H \to H$. Suppose that the adjoint operator A^* exists as an operator: $H \to H$, i.e.,

$$\langle A\Psi|\Psi'\rangle = \langle\Psi|A^*\Psi'\rangle \quad \text{for} \quad \Psi,\Psi' \in H.$$

A can then be extended to a linear transformation

$\underset{\sim}{D} \to \underset{\sim}{D}$ by the rule:

$$\langle A\alpha | \Psi \rangle = \langle \alpha | A^*\Psi \rangle \quad \text{for } \alpha \in \underset{\sim}{D}, \Psi \in H.$$

It is readily verified that this method of extension is natural in the sense that all algebraic operations are preserved.

This enables us to talk about the eigenvectors of A that may be in $\underset{\sim}{D}$ as "improper" eigenvectors. We can formulate the idea of "an orthonormal basis for H dependent on continuous parameters." Let M be a space, whose points we denote by p. Let dp be a measure on M. Consider mapping $p \to \alpha_p$ of M into $\underset{\sim}{D}$. It defines an *orthonormal basis* for H if:

$$\langle \Psi | \Psi' \rangle = \int \langle \Psi | \alpha_p \rangle \langle \alpha_p | \Psi' \rangle \, dp \quad \text{for } \Psi, \Psi \in H. \tag{2.1}$$

If, in addition, each such α_p is an eigenvector for A, i.e.,

$$A(\alpha_p) = a(p)\alpha_p \quad \text{for} \quad p \in M,$$

then the family $p \to \alpha_p$ may be thought of as the "eigenvector decomposition" for the operator A.

Suppose now that M is a complex manifold;

we may then speak of the mapping $p \to \alpha_p$ as being holomorphic (meromorphic) if, for each $\Psi \in H$, the complex-valued function $p \to \alpha_p(\Psi)$ on M is holomorphic (meromorphic).

We will be aiming to combine these two notions. Suppose $p \to \alpha_p$ is such a meromorphic map: $M \to \underset{\sim}{D}$. Suppose M_o is a real submanifold of M, and that (2.1) is true, where the integration is over M_o. Symbolically, we have:

$$\Psi = \int_{M_o} <\Psi|\alpha_p> \alpha_p \, dp. \qquad (2.2)$$

If the right-hand side of (2.2) is meromorphic, we may attempt to shift the contour of integration M_o, picking up the residues of the poles on the right-hand side as the contour passes through the poles. This device converts a continuous integral of Hilbert spaces into an asymptotic sum of discrete contributions.

EXAMPLE. (See Gelfand and Shilov [5].) Let H consist of the C^∞ functions $\Psi(x)$ of compact support defined over $-\infty < x < \infty$. For $Re(p) < -1$, x_+^p is defined as an element of $\underset{\sim}{D}$ by the formula

$$\langle x_+^p | \Psi \rangle = \int_0^\infty x^p \Psi(x) \, dx.$$

Let us define x_+^p over the whole complex p plane by using the relation:

$$\frac{d}{dx} x^p = p x^{p-1}.$$

Define $A: H \to H$ as follows:

$$A\Psi = \frac{d\Psi}{dx}.$$

Notice that $A^*(\Psi) = -d\Psi/dx$. Hence, <u>define</u> x_+^p for Re $p > -2$ as follows:

$$\langle x_+^p | \Psi \rangle = -\frac{1}{p+1} \langle x_+^{p+1} | \frac{d\Psi}{dx} \rangle,$$

i.e.,

$$x_+^p = \frac{1}{p+1} A x_+^{p+1}.$$

Continue on this way to define x_+^p, with poles at the integers $p = -1, -2, \ldots$. Notice also that:

$$x \frac{d}{dx} (x_+^p) = p x_+^p,$$

i.e., the x_+^p are vectors of the operator

$A' = x \frac{d}{dx}$. (Proof: By analytic continuation. It is true for Re p > -1, classically, hence true everywhere.)

In the following sections, we will describe this "analytic continuation" of eigenvectors of operators in cases that are more interesting for physical and group-theoretical investigations.

3. REDUCTION FROM SO(2,1) TO SO(1,1)

As a model for further work, let us consider the following situation. (See [8], part 1, for a fuller description.) Let H consist of the C^∞, complex-valued, periodic (of period 2π) function $\Psi(\theta)$, with:

$$<\Psi|\Psi> = \int_0^{2\pi} \Psi(\theta)^* \Psi'(\theta) \, d\theta.$$

Define operators as follows:

$$Z = \frac{d}{d\theta} \, ; \quad X_\lambda = \sin\theta \frac{d}{d\theta} + \lambda \cos\theta,$$

$$Y_\lambda = \cos\theta \frac{d}{d\theta} - \lambda \sin\theta.$$

Then, they satisfy the commutation relations of

the Lie algebra of SO(2,1) (or SL(2,R)). We will be interested in computing the eigenvectors of X_λ acting in D. Now we have:

$$X_\lambda = (\sin \theta)^{-\lambda}(\sin \theta \, \tfrac{d}{d\theta})(\sin \theta)^{\lambda}.$$

Let us change variables:

$$\Psi = \frac{i(e^{i\theta} - 1)}{e^{i\theta} + 1}.$$

Then, X_λ becomes

$$X_\lambda = \left(\frac{2x}{1+x^2}\right)^{-\lambda} \times \frac{d}{dx} \left(\frac{2x}{1+x^2}\right)^{\lambda},$$

$$d\theta = \frac{2\,dx}{1+x^2}.$$

Now, H can also be realized as the C^∞ functions $\Psi(x)$, $-\infty < x < \infty$, such that

$$\Psi(\infty) = \lim_{x \to \infty} \Psi(x) = \Psi(-\infty) = \lim_{x \to -\infty} \Psi(x),$$

and which are "differentiable at infinity" in the obvious sense (i.e., C^∞ differentiable at $y = 0$ after the change of variable $y = 1/x$). The inner product is

$$\langle \Psi | \Psi \rangle = \int_{-\infty}^{\infty} \frac{\Psi(x)^* \Psi(x) \, dx}{1 + x^2} \, .$$

Now, define Ψ_σ^+ as an element of $\underset{\sim}{D}$ as follows: Adopt the formula

$$\langle \Psi | \Psi_\sigma^+ \rangle^* = \int_0^1 \Psi(x) x^\sigma \left(\frac{2x}{1+x^2}\right)^{-\lambda} \frac{dx}{1+x^2}$$

$$+ \int_1^\infty \Psi(x) x^\sigma \left(\frac{2x}{1+x^2}\right)^{-\lambda} \frac{dx}{1+x^2} \, , \quad (3.1)$$

where this makes sense, i.e. for Ψ's which vanish to an appropriate order at 0 and infinity. Our job is to extend this over all of H by regularizing any possible divergent integrals.

Define $\Psi_\sigma^{+'}$ and $\Psi_\sigma^{+''}$ as follows:

$$\langle \Psi | \Psi_\sigma^{+'} \rangle = \int_0^1 \Psi(x)^* x^\sigma \left(\frac{2x}{1+x^2}\right)^{-\lambda} \frac{dx}{1+x^2} \, , \quad (3.2)$$

$$\langle \Psi | \Psi_\sigma^{+''} \rangle = \int_1^\infty \Psi(x)^* x^\sigma \left(\frac{2x}{1+x^2}\right)^{-\lambda} \frac{dx}{1+x^2} \, . \quad (3.3)$$

Notice that (3.2) makes sense for any $\Psi(x)$ if the $(\sigma - \lambda) > -1$, whereas (3.3) makes sense for any $\Psi(x)$, if $\text{Re}(\sigma - 3\lambda) < 1$. These two regions have points in common if $\lambda > -1$.

Following the regularization idea, let us rewrite (3.2) as:

$$<\Psi|\Psi_\sigma^{+\prime}> = 2^\sigma \left(\int_0^{1/2} \frac{x^{\sigma-\lambda}}{(1+x^2)^{-\lambda+1}} (\Psi(x)^* - \Psi(0)^*) \, dx + \Psi(0)^* \int_{1/2}^1 \frac{x^{\sigma-\lambda}}{(1+x^2)^{-\lambda+1}} \, dx + \Psi(0)^* \int_0^{1/2} \frac{x^{\sigma-\lambda}}{(1+x^2)^{-\lambda+1}} \, dx \right).$$

This is now well-defined and holomorphic in σ for $\text{Re}(\sigma - \lambda) > -2$, except for a pole at $\sigma - \lambda = -1$, coming from the third integral on the right-hand side. We can now continue this process, defining the left-hand side of (3.2) over the entire σ-plane, except for poles at $\sigma = \lambda -1$, $\lambda - 2$, ...

We can similarly deal with (3.3). Let us change variables:

MEROMORPHIC DECOMPOSITIONS

$y = 1/x$,

$$<\Psi | \Psi_\sigma^{+"}> = \int_0^1 \Psi(\tfrac{1}{y})^* \, y^{-\sigma} \, y^\lambda \left(\frac{1+y^2}{y^2}\right)^{\lambda+1} dy,$$

at $y = 0$, this integral has a singularity like

$$y^{-\sigma-\lambda-1}.$$

The poles obtained by regularizing this integral occur at the points:

$$-\sigma - \lambda - 1 = -1, -2, \ldots,$$

or

$$\sigma = -\lambda, -\lambda + 1, \ldots .$$

Thus, the poles of the meromorphic family of elements Ψ_σ^+ of D occur at

$$\sigma = \lambda - 1, \lambda - 2, \ldots, -\lambda, -\lambda + 1, -\lambda + 2, \ldots .$$

One can similarly define Ψ_σ^- as eigenvectors of X_λ by regularizing the integral over $-\infty < x \leq 0$. (The point is that SO(1,1) acting on the unit circle, in the complex plane, which is a "boundary homogeneous space" [7] of SO(2,1), has two orbits which have non-zero measures, namely the upper half circle and the lower half circle. This explains the "doubling" effect on the eigenvectors of the generator of its Lie algebra.)

4. ASYMPTOTIC EXPANSIONS OBTAINED FROM SPECTRAL RESOLUTIONS

As we have seen in the last section, the Dirac-type eigenvectors of the generator of the Lie algebra of the SO(1,1) subgroup of SO(2,1) have interesting analyticity properties. Let us now investigate in generality the consequences of these properties.

Suppose then that H is an incomplete Hilbert space and that $\tilde{D}$ is its Dirac space. Suppose A: H → H is an operator, such that its adjoint A* also exists as an operator: H → H, enabling us to extend A to an operator: $\tilde{D} \to \tilde{D}$.

Consider a one (complex) parameter family $\sigma \to \Psi_\sigma$ of elements of $\tilde{D}$, which depends meromorphically on σ, such that:

$$A(\Psi_\sigma) = \sigma \Psi_\sigma.$$

Suppose that:

$$<\Psi|\Psi'> = \int_{a-i\infty}^{a+i\infty} <\Psi|\Psi_\sigma><\Psi_\sigma|\Psi> \, d\sigma \qquad (4.1)$$

where it is assumed that the poles of Ψ_σ do not lie on the line over which the integration takes place.

Now, in trying to deform the contour of integration in (4.1) in accordance with Cauchy's theorem, we encounter the difficulty that the integrand is not a holomorphic function. To fix this up, let us introduce some additional notions. Let $\underset{\sim}{D}^*$ be the space of *conjugate linear* functionals on H, i.e., an element $\alpha \in \underset{\sim}{D}^*$ is a real-linear map: $H \to C$ such that

$$\alpha(c\Psi) = c^*\alpha(\Psi) \quad \text{for} \quad c \in C, \Psi \in H.$$

Write $\alpha(\Psi)$ as $<\Psi|\alpha>$. Define $<\alpha|\Psi>^*$.

DEFINITION. Let M be a space, whose points we denote by p, and with a measure denoted by dp.

Suppose $p \to \Psi_p \in \underset{\sim}{D}$, $p \to \Psi_p' \in \underset{\sim}{D}^*$ are families of elements of $\underset{\sim}{D}$ parametrized by points of M. They are said to define a *dual basis* for $\underset{\sim}{D}$ and $\underset{\sim}{D}^*$ if

$$<\Psi|\Psi'> = \int <\Psi|\Psi_p'><\Psi_p|\Psi'> \, dp \qquad (4.2)$$

for $\Psi, \Psi' \in H$.

EXAMPLE. Suppose H consists of the function $\Psi(x)$ of one variable with compact support. Define

$$<\Psi|\Psi'> = \int_{-\infty}^{\infty} \Psi(x)^* \Psi'(x) \, dx.$$

Let M consist of the real numbers. Define:

$$<\Psi_p|\Psi'> = \frac{1}{\sqrt{2\pi}} \int_{-\infty}^{\infty} e^{-ipx} \Psi'(x) \, dx;$$

$$<\Psi|\Psi_p'> = \frac{1}{\sqrt{2\pi}} \int_{-\infty}^{\infty} e^{ipx} \Psi(x)^* \, dx.$$

Then, the validity of (4.2) is just the Plancherel theorem for Fourier transforms. Particularly important for our purposes is the remark that both functions $p \to <\Psi_p|\Psi'>$ and $p \to <\Psi|\Psi_p'>$ are holomorphic. Thus, the integrand in (4.2) is now holomorphic, hence the contour of integration can be deformed using Cauchy's theorem.

Let us return to the general setting. Suppose H is a Hilbert space, σ a complex variable, and $\sigma \to \Psi_\sigma \in \underset{\sim}{D}$, $\sigma \to \Psi_\sigma' \in \underset{\sim}{D}^*$, mappings that depend meromorphically on σ. Suppose that

$$\langle \Psi | \Psi' \rangle = \int_{a-i\infty}^{a+i\infty} \langle \Psi | \Psi_\sigma' \rangle \langle \Psi_\sigma | \Psi' \rangle \, d\sigma. \qquad (4.3)$$

Suppose now that H_o is a subspace of H that is dense with respect to the Hilbert space topology for H. Then, we can in a natural way regard elements of $\underset{\sim}{D}$ as linear functionals on H_o. Intuitively, H_o consists of the functions in H that are zero in a neighborhood of the singularity set of the "functions" Ψ_σ.) Suppose further that $\langle \Psi_\sigma | \Psi' \rangle$ is a *holomorphic* function of σ, for $\Psi' \in H_o$. Then, the contour of integration in (4.2) can be shifted using Cauchy's theorem, obtaining

$$\langle \Psi | \Psi' \rangle = \int_{b-j\infty}^{b+j\infty} \langle \Psi | \Psi_\sigma' \rangle \langle \Psi_\sigma | \Psi' \rangle \, d\sigma$$

$$+ \sum_{j=1}^{n} \langle \Psi | \theta_j \rangle \langle \Psi_{\sigma_j} | \Psi' \rangle, \qquad (4.3)$$

where $\sigma_1,\ldots,\sigma_n$ are the points between the two lines of integration at which the function $\sigma \to \Psi_\sigma'$ has poles, and θ_j are the elements of $\underset{\sim}{D}$ which represent the residue at these poles.

Now, set $U(t) = \exp(tA)$. Suppose that $A(H_o) \subset H_o$, $U(t)(H_o) \subset H_o$, $U(t)(H) \subset H$. Suppose, as before, that

$$A(\Psi_\sigma) = \sigma\Psi.$$

Then

$$U(t)(\Psi_\sigma) = e^{t\sigma}\Psi_\sigma,$$

and

$$U(t)(\theta_j) = e^{t\sigma_j}\theta_j, \quad j = 1, 2, \ldots, n.$$

For $\Psi \in H$, $\Psi' \in H_o$,

$$\langle\Psi|U(t)\Psi'\rangle = \int_{b-i\infty}^{b+i\infty} \langle\Psi|\Psi_\sigma'\rangle\langle U(t)^*\Psi_\sigma|\Psi'\rangle \, d\sigma$$

$$+ \sum \langle\Psi|\theta_j\rangle\langle U(t)^*\Psi_{\sigma_j}|\Psi'\rangle$$

$$= \int_{b-i\infty}^{b+i\infty} e^{t\sigma} \langle\Psi|\Psi_\sigma'\rangle\langle\Psi_\sigma|\Psi'\rangle \, d\sigma$$

$$+ \sum e^{t\sigma_j} \langle\Psi|\theta_j\rangle\langle\Psi_{\sigma_j}|\Psi'\rangle \, . \qquad (4.4)$$

This gives us a representation for the asymptotic behavior of the matrix elements $\langle\Psi|U(t)|\Psi'\rangle$ as $t \to \infty$, assuming, of course, that if b is sufficiently negative, the first term involving integration can be ignored.

In turn, (4.3) suggests a group-theoretic interpretation of these facts. Regard H and $\underset{\sim}{D}$ as spaces of linear functionals on H_0. $t \to U(t)$ defines a one-parameter group of transformations on H, i.e., a linear representation of the additive group R of real numbers. Equation (4.3) can be written in the form:

$$\Psi = \int_{a-i\infty}^{a+i\infty} \langle\Psi|\Psi_\sigma'\rangle \Psi_\sigma \, d\sigma,$$

and can be interpreted as decomposing D into a "direct integral" of subspaces in each of which $t \to U(t)$ acts irreducibly. On the other hand,

(4.3) can be described symbolically as:

$$\Psi \sim \langle\Psi|\theta_1\rangle \Psi_{\sigma_1} + \ldots \tag{4.5}$$

and may be interpreted as a "decomposition" of $\underset{\sim}{D}$ into a discrete "asymptotic" sum of subspaces in each of which the group acts irreducibly. In this form, the idea can be generalized to other groups, as we shall indicate in the next section.

5. A POSSIBLE GENERALIZATION FOR SUBGROUP DECOMPOSITIONS

We will now present one method of implementing the decomposition suggested by (4.5) for other more complicated types of group representations. Suppose that D is a representation of a Lie group G on a Hilbert space H. Let D also denote the infinitesimal representation of the Lie algebra $\underset{\sim}{G}$.

Suppose L is a subgroup of G, with its Lie algebra $\underset{\sim}{L}$ then a subalgebra of $\underset{\sim}{G}$. Let $U(\underset{\sim}{G})$ be the universal enveloping algebra of G, with $U(\underset{\sim}{L})$ a subalgebra of $U(\underset{\sim}{G})$. Suppose $\Delta_1,\ldots,\Delta_r$ are elements of U($\underset{\sim}{L}$) that form a maximal abelian subset. Then, one can consider $D(\Delta_1), \ldots, D(\Delta_r)$ as operators on H. Suppose the operators of U($\underset{\sim}{L}$) are defined over all of H, and their adjoints are defined over all of H. Then, the operators $D(\Delta_1), \ldots, D(\Delta_r)$ and their adjoints are defined over all of H, where they can be extended to be operators on $\underset{\sim}{D}$, the Dirac space associated with H. Since they commute, there are no algebraic obstacles to simultaneously diagonalizing them. Suppose this can be done. The eigenvectors then depend on r parameters, $\sigma_1, \ldots, \sigma_r$, the eigenvalues. Just as before, we can formulate what it means for these eigenvectors (elements of $\underset{\sim}{D}$) to depend holomorphically or meromorphically on the complex parameters, and what it would mean to replace integrals of the form:

$$<\Psi|\Psi'> = \int_{a_1-i\infty}^{a_1+i\infty} \cdots \int_{a_r-i\infty}^{a_r+i\infty} <\Psi|\Psi_\sigma'><\Psi_\sigma|\Psi'> d\sigma_1 \ldots d\sigma_r$$

by discrete sums via the calculus of residues. This seems to be the meaning of the interesting calculations done by Sciarrino and Toller [11], for the case G = SO(3,1), H = SO(2,1), which in turn forms the mathematical foundation for Toller's group-theoretic treatment of the "daughter trajectory" phenomenon [13]. We will pursue this case further in the next section.

If L is semisimple, the simplest and most useful choice of $\Delta_1,\ldots,\Delta_r$ seems to be the following:

Let $\underset{\sim}{L} = \underset{\sim}{K} + \underset{\sim}{P}$ be the Cartan decomposition, [7], i.e. $\underset{\sim}{K}$ is a maximal compact subalgebra, $[\underset{\sim}{K}, \underset{\sim}{P}] \subset \underset{\sim}{P}$, $[\underset{\sim}{P}, \underset{\sim}{P}] \subset \underset{\sim}{K}$. Let $\underset{\sim}{A}$ be a maximal abelian subalgebra of $\underset{\sim}{P}$, and let $\underset{\sim}{B}$ be a maximal abelian subalgebra of $\underset{\sim}{K}$ that commutes with $\underset{\sim}{A}$. Then, $\underset{\sim}{C} = \underset{\sim}{A} + \underset{\sim}{B}$ is a maximal abelian (and Cartan) subalgebra of $\underset{\sim}{L}$. Choose $\Delta_1,\ldots,\Delta_s$, as a basis of $\underset{\sim}{C}$. Now, choose $\Delta_{s+1},\ldots,\Delta_r$ as a basis for the center (i.e. the Casimir operators) of $U(\underset{\sim}{L})$. (In fact, one can prove that s = r/2). Thus, the eigenvalues $\sigma_{s+1},\ldots,\sigma_r$ would label the representation, while the $\sigma_1,\ldots,\sigma_s$ would label the eigenvalues of the elements of $\underset{\sim}{L}$ (since L = KAK, a generalized "Euler angle" decomposition) and would determine the asymptotic behavior of the matrix elements of the elements of L.

In order to fix these ideas, let us turn to the case $G = SL(2,C)$, $L = SL(2,R)$ (or, equivalently, $G = SO(3,1)$, $L = SO(2,1)$) which is relevant to the "daughter trajectory" phenomenon.

6. REDUCTION FROM SL(2,C) TO SL(2,R)

Let H be the space of complex-valued functions (not necessarily holomorphic) $\Psi(z,z^*)$ that are defined on the complex z-plane. For the moment we will suppose them to be C^∞ and of compact support, so we can define the inner product as follows:

$$<\Psi|\Psi'> = \int \Psi(z,z^*)^* \Psi'(z,z^*) \, dz \, dz^*.$$

Let G be the group $SL(2,C)$ of 2×2 complex matrices of determinant one. Define a representation of G in H (following Naimark [10]) as follows:

$$D(g)(\Psi)(z) = |g_{12} z + g_{22}|^\lambda (g_{12} z + g_{22})^m \Psi(g^T z),$$

$$\text{for} \quad g \in G, \; \Psi \in H,$$

$$g = \begin{pmatrix} g_{11} & g_{12} \\ g_{21} & g_{22} \end{pmatrix}$$

λ is a complex number, m an integer. They may be considered as the parameters for the representation

$$g^T z = \frac{g_{11} \; z + g_{21}}{g_{12} \; z + g_{22}} .$$

Consider the subgroup $L = SL(2,R)$, i.e. the matrices in G with real coefficients. Then, L acting on 3-space has three orbits, the upper and lower half planes, and the real axis. The irreducible representations of L are obtained by letting it act on the real axis, as follows:

Let H' consist of the functions $\theta(x)$ ($x = \frac{1}{2}(z+z^*)$), with

$$\langle \theta | \theta' \rangle = \int \theta(x)^* \theta'(x) \, dx.$$

Define a representation D' of L on H' as follows:

$$D'(g)(\theta)(x) = |g_{12} \, x + g_{22}|^\alpha (g_{12} \, x + g_{22})^\beta \theta(g^T x) dx$$

α is a complex parameter, β is zero or -1.

Our goal is to find the eigenvectors in $\underset{\sim}{D}$

MEROMORPHIC DECOMPOSITIONS 23

(the Dirac space associated with H) corresponding to a maximal abelian set in $U(\underset{\sim}{L})$ by constructing intertwining operators: $H' \to H$, then extending them to the Dirac space. In fact, we will construct these intertwining operators as integral operators of the form:

$$\Psi(z) = \int k(z,x)\theta(x)\,dx.$$

To construct the kernel, k, let us do some calculations: Suppose $g \in SL(2,R)$.

$$g^T z - g^T z^*$$

$$= \frac{(g_{11}z+g_{21})(g_{12}z^*+g_{22}) - (g_{11}z^*+g_{21})(g_{12}z+g_{22})}{(g_{12}z+g_{22})(g_{12}z^*+g_{22})}$$

$$= \frac{z-z^*}{(g_{12}z+g_{22})(g_{12}z^*+g_{22})}$$

Similarly,

$$g^T z - g^T x = \frac{z-x}{(g_{12}z+g_{22})(g_{12}x+g_{22})}$$

Also,

$$d(g^T x) = \frac{dx}{|g_{12} z + g_{22}|}$$

Suppose we require that

$$D(g)(\Psi)(z) = \int k(z,x) D'(g) \theta(x)\, dx \qquad (6.1)$$

for $g \in SL(2,R)$.

The right hand side of (6.1) is

$$\int k(z,x) |g_{12} x + g_{22}|^\alpha (g_{12} x + g_{22})^\beta \theta(g^T x)\, dx$$

Change variables in the integral, $x' = g^T x$, obtaining:

$$\int \frac{k(z, g^{T-1} x')}{k(g^T z, x')} \frac{k(g^T z, x') |g_{12} x + g_{22}|^\alpha}{|g_{12} x + g_{22}|^{-1}}$$

$$(g_{12} x + g_{22})^\beta \theta(x')\, dx'$$

The left hand side of (6.1) is now:

$$\int |g_{12} z + g_{22}|^\lambda (g_{12} z + g_{22})^m k(g^T z, x') \theta(x')\, dx'.$$

We must then have:

MEROMORPHIC DECOMPOSITIONS

$$\frac{k(z, g^{T-1}x')}{k(g^T z, x')} |g_{12}x+g_{22}|^{\alpha+1}(g_{12}x+g_{22})^{\beta}$$

$$= |g_{12}z+g_{22}|^{\lambda}(g_{12}z+g_{22})^{m}$$

Now

$$\frac{k(z, g^{T-1}x')}{k(g^T z, x')} = \frac{k(z, gx)}{k(g^T z, g^T x)}$$

This suggests a possible form for k:

$$k(z, x) = |z+z^*|^a |z-x|^b (z-x)^c \qquad (6.2)$$

Then

$$\frac{k(z, x)}{k(g^T z, g^T x)}$$

$$= |(g_{12}z+g_{22})(g_{12}x+g_{22})|^b [(g_{12}z+g_{22})(g_{12}x+g_{22})]^c$$

$$|g_{12}z+g_{22}|^{2a}$$

Hence

$$|(g_{12}z+g_{22})(g_{12}x+g_{22})|^b |(g_{12}z+g_{22})(g_{12}x+g_{22})|^c$$

$$|g_{12}z+g_{22}|^{2a}|g_{12}x+g_{22}|^{\alpha+1}(g_{12}x+g_{22})^{\beta}$$

$$= |g_{12}z+g_{22}|^{\lambda}(g_{12}z+g_{22})^{m}$$

First of all, we see that:

$c = m$

$2a+b = \lambda$ \hfill (6.2)

This takes care of the terms involving z. Let us equate the terms involving x.

Case 1. $c = m$ is even.

We see that β must be zero, since $(g_{12}x+g_{22})^{c} = |g_{12}x+g_{22}|^{c}$. Hence,

$b + m + \alpha + 1 = 0$ \hfill (6.3)

Case 2. $c = m$ is odd, i.e. $c = 2n+1$, with n an integer.

Then, $\beta = -1$.

$b + 2n + \alpha + 1 = 0$, or

$b + m + \alpha = 0$ \hfill (6.4)

We can now sum up what we have proved in the following way:

THEOREM 6.1. Suppose $H^{\lambda,m}$ and $H^{\alpha,\beta}$ are function spaces, consisting respectively of functions of the form $\Psi(z,z^*)$ and $\theta(x)$, on which $SL(2,C)$ and $SL(2,R)$, respectively, act according to representations labelled by (λ, m) and (α, β), respectively. Then there is a "formal" integral-operator $H^{\alpha,\beta} \to H^{\lambda,m}$ intertwining the action of $SL(2,R)$ on both spaces, of the form,

$$\Psi(z,z^*) = \int_{-\infty}^{\infty} |z-z^*|^a |z-x|^b (z-x)^m \theta(x) \, dx, \quad (6.5)$$

where a and b are determined by (6.2) - (6.4).

Now, let us proceed to sketch the analysis of Sciarrino and Toller using the theory of generalized functions. Consider a matrix element of the form:

$$\langle\Psi'|\Psi\rangle = \iiint |z+z^*|^a |z-x|^b (z-x)^m \Psi'(z,z^*)^* \theta(x) \, dx \, dz \, dz^* \quad (6.6)$$

For certain values of a, b and m, and/or

certain choices of θ and Ψ' in their respective Hilbert spaces, the integrals may diverge. For example, let us consider this matrix element for Ψ' and θ that belong to subspaces that transform according to an irreducible (hence finite dimensional) representation of the maximal compact subgroups, SU(2) and SO(2,R). For this choice, we know automatically that there will be no singularities arising from the singularity at infinity. (For, the integrals can be alternately considered as integrals over compact spaces, namely the 2-sphere and the circle, by stereographic projection, and Ψ' and θ then become functions that are everywhere continuous.) We must then only worry about the integral for z near the real axis. (6.5) will then converge for a certain range of values of a, b and m. The problem is one of analytic continuation to the other values. The subtraction procedure described in great detail by Gelfand and Shilov [5] works here to prove it is a meromorphic function, with poles separated by integers. We will not carry out the details: Obviously we would only be duplicating the work of Sciarrino and Toller, who use another method. (They show that the integral is given by a hypergeometric function, where poles can directly be computed.) The method described here might be useful in establishing general properties of these poles and their residues, and in pointing the way towards extensions to other groups. In fact, we will now turn to a general description of what we have done here.

7. A GENERAL FORMULATION OF THE REDUCTION PROCEDURE

Suppose G is a group, L a subgroup, σ a representation of L by operators in a vector space V. σ determines a vector bundle E on the homogeneous space G/L, whose fibre over the identity coset is V. Let H be the vector space of cross-sections of the vector bundle E. Then, G acts on these cross-sections, determining a representation D of G in H. This is the representation of G *induced* by the representation σ of L.

Now, suppose that G' is another subgroup of G, and our problem is to reduce the representation

D restricted to G' into irreducible representations. The problem now separates into two parts:

a) Find the space of orbits of G' acting on G/L. Decompose each cross-section of E into an integral over the orbit space of cross-sections over the individual orbits.

b) Suppose one of the orbits of G' on G/L is of the form G'/L', where L' is a subgroup of G'. The problem now is to decompose the action of G' on the cross-sections of the vector bundle E restricted to G'/L' into irreducible representations.

Problem a) has been described in general terms in [7]. Here, we will make some comments about problem b), and its connection with the preceding work.

We are then given a representation σ of L' by operators on a vector space V, determining a vector bundle as E' on G'/L'. The space of cross-sections on this bundle can be realized as the space of mappings Ψ: G' $\to$ V satisfying:

$$\Psi(g\ell) = \sigma(\ell^{-1})\Psi(g) \quad \text{for} \quad \ell \in L'.$$

MEROMORPHIC DECOMPOSITIONS   31

The action of G' on cross-sections is, in this
presentation, just left-translation.

Now, suppose that S is another subgroup of
G', and that α is a representation of S by linear
transformations on V. Suppose that $g \to \Psi(g)$ is a
mapping from G to V. Assign to Ψ the mapping

$$g \to \theta(g) = \int_S \alpha(s)\Psi(gs)\, ds \qquad (7.1)$$

where ds is a left-invariant volume element on S.
Then,

$$\theta(gs) = \alpha(s^{-1})\theta(g) \quad \text{for} \quad g \in G',\ s \in S.$$

Thus, (7.1) defines a "formal" (since the integral
(7.1) might not converge) intertwining mapping of
the space of cross-sections of the vector bundle
determined by σ on the coset space G'/L' into the
space of cross-sections of the vector bundle on
G'/S determined by α.

Now, suppose that V is a complex-vector
space with a Hermitian inner product (,) which
is invariant under the action of α. Here, if θ
and θ' represent two cross sections of the vector
bundle on G/S determined by α,

$$g \to (\theta(g), \theta'(g))$$

is a complex-valued function which is invariant under right-translation by S, i.e. defines a function on G/S. Define the inner product on cross-sections as:

$$\langle\theta|\theta'\rangle = \int_{G'/S} (\theta(g), \theta'(g))\, dp,$$

where "dp" is a volume element on the coset space G'/S. Let H be the Hilbert space so obtained.

Then, if $g \to \Psi(g)$ represents a cross-section of the vector bundle on G'/L' determined by σ, then

$$\theta' \to \langle\theta|\theta'\rangle,$$

where θ is given by (7.1), determines a linear functional on H, provided, of course, that the integrals converge. If not, the functional can be determined by one of the methods of regularization, [5], e.g. by "analytic continuation" relative to the representations σ, α used to define the two bundles, or by some subtraction procedure. What is clearly needed is some sort of geometric-group theoretic method of performing these regularizations.

For example, suppose we look at the following situation: G' is a semisimple, connected noncompact Lie group with a finite center, $S = K$, a maximal compact subgroup. Let $\underline{G}' = K \oplus \underline{P}$ be its Cartan decomposition, i.e. $[\underline{K}, \underline{P}] \subset \underline{P}$, $[\underline{P}, \underline{P}] \subset \underline{K}$. Let X_o be an element of $\underline{P}$, $C(\underline{X}_o)$ the centralizer of X_o in $\underline{G}'$, $\underline{N}^+(X_o)$ the nilpotent subalgebra spanned by the eigenvectors if Ad X_o with positive eigenvalues. (Recall that Ad X_o is diagonalizable, and has real eigenvalues.) Let $C(X_o)$ be the centralizer of X_o in G', and let $N^+(X_o)$ be the connected subgroup of G' generated by the algebra $\underline{N}^+(X_o)$. Let $L' = C(X_o)N^+(X_o)$.

We can reduce (7.1) to a more recognizable form. Suppose S' is the subgroup $L_r'S$. Suppose q denotes a point of the coset space S/S'. Suppose ds' is a left-invariant volume element on S'. If $s \to f(s)$ is a function on S, define $\hat{f}(q)$ as follows:

$$s \to \int_{S'} f(ss') \, ds'$$

is a function on S which is invariant under right multiplication by S', hence only depends on q, i.e.,

$$\hat{f}(q) = \int_{S'} f(ss') \, ds'.$$

Suppose dq is a volume element on S/S' such that

$$\int \hat{f}(q) \, dq = \int_S f(s) \, ds.$$

Symbolically, one can write this as:

$$\int_S f(s) \, ds = \int_{S/S'} d(s/s') \left(\int_{S'} f(ss') \, ds' \right)$$

Thus, the right hand side of (7.1) can be written as:

$$\int_{S/S'} d(s/s') \left(\int_{S'} \alpha(ss') \Psi(gss') \, ds' \right)$$

$$= \int_{S/S'} d(s/s') \left(\int_{S'} \alpha(s) \alpha(s') \sigma(s'^{-1}) \Psi(gs) \, ds' \right)$$

For example, in the case we treated above $G' = SL(2,R)$, $S = SO(2,R)$, $S' = $ identity, $L' = $ the subgroup such that G'/L' is the circle in the complex plane. G'/S is then the interior of the unit circle. By a "Cayley transform" this can be transformed to the case where G'/L' is the real axis in the complex plane, and G'/S is the upper (or lower) half plane.

8. LIMITS AND CONTRACTIONS OF LIE GROUPS AND THE $t = 0$ SYMMETRY

In [8, part 5] we pointed out that there is a geometric pattern to the current investigations in the physics literature concerning the consistency of the partial wave and Regge expansions of the scattering amplitude near $t = 0$, i.e. the point where the "little group" changes from compact to non-compact. In this section we will pursue this investigation, particularly pointing out certain connections with the theory of limits and contractions of groups [7, 8].

Let us consider a typical problem of this type. Suppose given a scattering process whereby two initial particles, of four-momenta p_1, p_2 go into particles of momenta p_3, p_4. For simplicity, suppose the particles are spinless, with masses m_j^2, $j = 1, 2, 3, 4$, i.e.

$$p_j^2 = m_j^2, \quad j = 1, 2, 3, 4. \tag{8.1}$$

(p^2 = Lorentz metric square of $p \in R^4$).

The amplitude f is then a function of (p_1, p_2, p_3, p_4) of these vectors; f is only defined, however, on sets of p's constrained by the conservation of momentum:

$$p_1 + p_2 = p_3 + p_4 \tag{8.2}$$

Let us follow Toller's method [12] for describing the "harmonic analysis" of f. Introduce the invariants, the "momentum transfer"

$$t = (p_1 - p_3)^2$$

and "center of mass energy"

$$s = (p_1 + p_2)^2.$$

Also, set $q = p_1 - p_3$. Notice that the subset of R^{16} defined as the set of p_1, p_2, p_3, p_4 satisfying (8.1), (8.2) with t a fixed value, is acted on by a subgroup of SO(3,1), the Lorentz group. In fact, let L(q) be the set of $g \in SO(3,1)$ such that

$$gq = q.$$

$L(q)$ is then the "little group" of q. As is well known, it is isomorphic to $SO(3,R)$ if $t = q^2 > 0$, to $SO(2,1)$ if $t < 0$, to E_2, the group of rigid motions in the plane, if $t = 0$, but $q \neq 0$, to $SO(3,1)$ itself if $q = 0$. In the papers by Toller and coworkers one will find the most extensive discussion of the application of these remarks to problems of physical interest.

At any rate, let $N(q)$ be the set of four-vectors (p_1, p_2, p_3, p_4) satisfying (8.1) - (8.2), with q fixed G g ε $L(q)$ acts on $N(q)$ as follows:

$$g(p_1, p_2, p_3, p_4) = (p_1, gp_2, p_3, gp_4).$$

One sees readily that $L(q)$ acts transitively on $N(q)$. If $(p_1, p_2, p_3, p_4) \varepsilon N(q)$, the isotropy subgroup of $L(q)$ at this point, is the intersection of the isotropy subgroups of $SO(3,1)$ at p_3 and p_4.

This construction has a general interpretation: Let G be the Lorentz group $SO(3,1)$. Let G act on R^{16} as follows:

$$g(p_1, p_2, p_3, p_4) = (p_1, gp_2, p_3, gp_4).$$

Notice now that $L(q)$ is a subgroup of G that maps $N(q)$ into itself.

Suppose now that $q_1, q_2, \ldots$ is a sequence of four-vectors converging to a four-vector q.

THEOREM 8.1. The "limit" of the subgroups $L(q_1)$, $L(q_2), \ldots$ (as defined in [7], Chapter 11) is contained in the subgroup $L(q)$ of G.

Proof: Let us recall the definition of "limit" of a sequence $L_1, L_2, \ldots$ of subgroups of a typological group G. We say that

$$\lim_{n \to \infty} L_n \subset L, \qquad (8.3)$$

if, *whenever* a sequence $\ell_1, \ell_2, \ldots$, each $\ell_n \in L_n$, converges to an $\ell \in G$, this limit lies in L.

Suppose now that $P^1 = (p_1^1, p_2^1, p_3^1, p_4^1)$, $P^2 = (p_1^2, p_2^2, p_3^2, p_4^2), \ldots$ is a sequence of points of R^{16}, with $q_n = p_1^n - p_3^n$, $n = 1, 2, \ldots$, each $P^n \in N(q_n)$, which converges to a point $P \in N(q)$. Thus, if $\ell_n \in L(q_n)$ converges to $\ell \in G$,

$$\ell_n P^n \to \ell P \quad \text{as} \quad n \to \infty$$

Hence, $\ell(N(q)) = N(q)$. Indeed, then, $\ell \in L(q)$, which is what is required to prove (8.3).

Suppose now that q_n^2 is either constantly positive or negative for all n. Then, it is well-known that the "little groups" $L(q_n)$ are all mutually conjugate within G, i.e. there is a sequence (g_n) of elements of $SO(3,1) = G$ such that:

$$L(q_n) = g_n L(q_j) g_n^{-1},$$

hence relation (8.3) takes the form:

$$\lim_{n \to \infty} g_n L(q_j) g_n^{-1} \subset L(q). \tag{8.4}$$

We shall now turn to the study of the effect of these limiting relations on the harmonic analysis of functions on the groups involved.

9. ANALYTIC CONTINUATION OF THE FOURIER EXPANSION OF FUNCTIONS ON $SO(3,R)$ AND $E(2)$

Let $\underset{\sim}{K}$ be the Lie algebra of $SO(3,R)$. Choose a basis (X_i), $1 \leq i, j, k \leq 3$, (summation convention) satisfying the structure relations:

$$[X_i, X_j] = \varepsilon_{ijk} X_k. \tag{9.1}$$

Define a "deformation" of this Lie algebra (see [8]) depending on a parameter $-\frac{1}{4} \leq t \leq \frac{1}{4}$, as follows:

$$[X_1, X_3]_t = [X_1, X_3]$$
$$[X_2, X_3]_t = [X_2, X_3]$$
$$[X_1, X_2]_t = 4t[X_1, X_2] \tag{9.2}$$

For $t = \frac{1}{4}$, this gives the original algebra (9.1). For $t = 0$, (9.2) defines the Lie algebra of $E(2)$, the group of rigid motions of the plane. For $t = -\frac{1}{4}$, it gives the Lie algebra of $SO(2,1)$. We will denote the tth Lie algebra structure by $\underset{\sim}{K}_t$.

For each t, let K_t be a Lie group whose Lie algebra is $\underset{\sim}{K}_t$. Let $\exp_t: \underset{\sim}{K}_t \to K_t$ be the exponential map defined for this algebra. Let α, β, γ be real variables. Introduce them as coordinates on K_t as follows:

$$(\alpha, \beta, \gamma) \to \exp_t(\alpha X_3) \exp_t(\beta X_1) \exp_t(\gamma X_3) \tag{9.3}$$

MEROMORPHIC DECOMPOSITIONS

Thus, α, β, γ are the "Euler angles" for each group K_t. A function $f(\alpha, \beta, \gamma, t)$ of the indicated variables thus determines via (9.3) a family $f_t(k_t)$ of functions, defined on each K_t. We will be interested in the variation of the Fourier expansion of this function on K_t with t.

We will need two further facts:

a) for $t > 0$, define A_t: $\underset{\sim}{K} \to \underset{\sim}{K}$ as follows:

$$A_t(X_3) = X_3, \quad A_t(X_j) = 2\sqrt{t}\, X_j, \quad j = 1, 2$$

Notice that the relation (9.2) can be summed up as follows:

$$[X_i, X_j]_t = A_t^{-1}[A_t X_i, A_t X_j] \qquad (9.4)$$

This shows that the algebras $\underset{\sim}{K}_t$ are mutually isomorphic. However, the isomorphism breaks down at $t = 0$, as it must. (This is just the "contraction" idea of Inonu and Wigner, of course.)

b) Let G be the Lorentz group SO(3,1), with $\underset{\sim}{G}$ its Lie algebra. The basis of $\underset{\sim}{G}$ can be chosen of the form (X_i, Y_j), with:

$$[X_i, X_j] = \varepsilon_{ijk} X_k$$
$$[X_i, Y_j] = \varepsilon_{ijk} Y_k$$
$$[Y_i, Y_j] = -\varepsilon_{ijk} X_k. \tag{9.5}$$

Notice that:

$$[X_1 + Y_2, X_2 - Y_1] = 0. \tag{9.6}$$

Define $\phi_t: \underline{K}_t \to \underline{G}$ as follows:

$$\phi_t(X_3) = X_3$$
$$\phi_t(X_1) = (t + 1)X_1 + (t - 1)Y_2$$
$$\phi_t(X_2) = (t + 1)X_2 - (t - 1)Y_1 \tag{9.7}$$

Notice that ϕ_t is a Lie algebra homomorphism, i.e. maps the structure relations (9.2) into (9.5). ϕ_t then also determines an isomorphism, which we also denote by ϕ_t, of K_t with a subgroup of G.

Now we must write down the Fourier expansion of a function $k_t \to f_t(k_t)$ on K_t, for $t \neq 0$. As is well-known, it takes the following form:

$$f_t(k_t) = \sum_{j=0}^{\infty} \text{trace}(F_t{}^j D_t{}^j(k_t)(2j + 1). \tag{9.8}$$

MEROMORPHIC DECOMPOSITIONS   43

$k_t \to D_t^j$ is the spin -j irreducible unitary representation of $SO(3,R)$.

$$F_t^j = \int_{K_t} f_t(k_t) D_t^j(k_t^{-1}) dk_t. \qquad (9.9)$$

dk_t is the invariant value element on K_t, with the value element normalized to one. Let J_t^2 be the second order Casimir operator of $\underset{\sim}{K}_t$. Then,

$$j(j+1) = D_t^j(J_t^2),$$

hence

$$2j + 1 = \sqrt{1 + 4D_t^j(J_t^2)}$$

(9.8) can be written as:

$$f_t(k_t) = \sum_{j=0}^{\infty} \text{trace}(F_t^j D_t^j \sqrt{1 + 4D_t^j(J_t^2)}) \qquad (9.10)$$

Now, suppose that $g \to D(g)$ is an irreducible unitary representation of $G = SO(3,1)$ on a Hilbert space H, which splits up under $J_t(K_t)$ ($t \neq 0$), into a direct sum of the spin j-representations, $j = 0, 1, \ldots$. (See Naimark [10] for a description of these representations that do this job.) Notice that (9.10) can then be written in the form:

$$f_t(k_t) = \text{trace}(F_t D(\phi_t(k_t)) \sqrt{1+4D\phi_t(J_t^2)}) \qquad (9.11)$$

with:

$$F_t = \int_{K_t} f_t(k_t) D(\phi_t(k_t^{-1})) \, dk_t \qquad (9.12)$$

(9.11) and (9.12) are now in a convenient form for carrying out the limit as $t \to 0$. Suppose $f_t(k_t)$ is given as a function $f(t, \alpha, \beta, \gamma)$, via (9.3). For $t = \frac{1}{4}$, the invariant value element is

$$\frac{1}{8\pi^2} \sin \beta \, d\alpha \, d\beta \, d\gamma.$$

The effect of A_t, which converts K into K_t, is the substitution:

$$\alpha \to \alpha, \; \gamma \to \gamma, \; \beta \to \frac{\beta}{2\sqrt{t}}$$

Here, (9.12) takes the form:

$$F_t = \frac{1}{8\pi^2} \int_0^\pi \int_0^{2\pi} \int_0^{2\pi} f_t(t, \alpha, \frac{\beta}{2\sqrt{t}}, \gamma)$$

$$D(\phi_t(k_t^{-1})) \sin\beta \, d\beta \, d\alpha \, d\gamma$$

$$= \frac{1}{8\pi^2} \int_0^{\pi/2\sqrt{t}} \int_0^{2\pi} \int_0^{2\pi} f(t, \alpha, \beta, \gamma)$$

$$D(\phi_t(\exp_t(\alpha X_3)\exp_t(-\beta X_1)\exp_t(-\alpha X_3)))$$

$$\frac{\sin(2\sqrt{t}\beta)}{2\sqrt{t}} \, 2t \, d\beta \, d\alpha \, d\gamma$$

Hence,

$$\frac{F_t}{t} \to \frac{1}{4\pi^2} \int_0^\infty \int_0^{2\pi} \int_0^{2\pi} f(0, \alpha, \beta, \gamma)$$

$$D(\phi_0(\exp_0(-\gamma X_3)\exp_0(-\beta X_1)\exp_0(-\alpha X_3))$$

$$\beta \, d\beta \, d\alpha \, d\gamma$$

as $t \to 0$. (9.13)

The integral on the right hand side of (9.13) is readily seen to be the invariant volume element on the group K_0, isomorphic to $E(2)$.

Let us calculate $D(\phi_t(J_t^2))$. Now,

$$\phi_t(J_t^2) = -\left(\phi_t(X_3)^2 + \left(\frac{\phi_t(X_1)}{2\sqrt{t}}\right)^2 \right.$$
$$\left. + \left(\frac{\phi_t(X_2)}{2\sqrt{t}}\right)^2\right)$$
$$= -\frac{1}{4t}(4t\phi_t(X_3)^2 + \phi_t(X_1)^2 + \phi_t(X_2)^2). \quad (9.14)$$

Then $\phi_t(J_t^2)t$ goes into, as $t \to 0$, the Casimir operator of $\underset{\sim}{K}_0$, $-\phi_0(X_1)^2 - \phi_0(X_2)^2$, which we denote by Δ_0. Suppose $f(k_0) = \lim_{t \to 0} f_t(k_t)$. From (9.11), (9.13), (9.14) we then have:

$$f(k_0) = \lim_{t \to 0} \sqrt{t} \; \text{trace}(F_0 D(\phi_t(k_t))\sqrt{\Delta_0}) \quad (9.15)$$

with

$$F_0 = \int_{K_0} f(k_0) D(\phi_0(k_0)) dk_0$$

We see that (9.15) expresses $f(k_0)$ f as a "sum" of matrix elements of unitary representation of K_0. Presumably, when one carries out the limit on the right hand side of (9.15) the Fourier expansion theorem for functions on $K_0 = E(2)$ is obtained, although we will not attempt the calculation at this stage.

MEROMORPHIC DECOMPOSITIONS

This work also suggests a conjectured form for the Fourier expansion of functions on $L = SO(2,1)$. Let ϕ_{-1} be a homomorphism: $K_{-1} \to SO(3,1)$. By analogy with (9.11), it might take the form:

$$f(k) = \text{trace}(FD(\phi(k))\sqrt{1 + 4D\phi_{-1}(\Delta_\ell)}) \quad (9.16)$$

with

$$F = \int_{SO(2,1)} f(k) D(\phi(k^{-1})) dk,$$

and with Δ the second-degree Casimir operator of $\underset{\sim}{K}_{-1}$. Again, we must defer study of this point, together with study of the possible applications to the scattering amplitude, to a later chapter.

BIBLIOGRAPHY

1. J. F. Boyce, International Centre for Theoretical Physics Preprint, IC/66/30, Trieste (1966).

2. G. Domokos and P. Suranyi, Nucl. Phys. *54*, 529 (1964).

3. D. Z. Freedman and J. M. Wang, Phys. Rev. Letters *17*, 569 (1966).

4. I. M. Gel'fand, M. I. Graev and N. Ya. Vilenkin, Generalized Functions, Vol. 5, Academic Press, New York (1966).

5. I. M. Gel'fand and G. E. Shilov, Generalized Functions, Vol. 1, Academic Press, New York (1964).

6. F. T. Hadjiannou, Nuovo Cimento *44*, 185 (1966).

7. R. Hermann, Lie Groups for Physicists, W. A. Benjamin, New York (1966).

8. _____. Analytic Continuation of Group Representations, Comm. Math. Phys., Part I, 2, 251-270 (1966); Part II, 3, 53-74 (1966); Part III, 3, 75-97 (1966).

9. H. Joos, Lectures in Theoretical Physics,

Vol. IIA, p. 132, edited by W. E. Brittin and A. O. Barut, Boulder (1965).

10. M. A. Naimark, Linear Representations of the Lorentz Group, Pergamon Press, London (1964).

11. A. Sciarrino and M. Toller, Internal Report No. 108, Istituto di Fisica, "G. Marconi," Roma (1966).

12. L. Sertorio and M. Toller, Nuovo Cimento *33*, 413 (1964).

13. M. Toller, An Expansion of the Scattering Amplitude at Vanishing Four-Momentum Transfer Using the Representations of the Lorentz Group, CERN report, 67/582/5 - TH 780, (1967).

CHAPTER II

THE FOURIER TRANSFORM ON LIE GROUPS

INTRODUCTION

In Chapter I we presented some problems in representation theory that arise in a natural way in the group-theoretic side of S-matrix theory. Now, to treat these problems in a systematic way will require the full force of the theory of Fourier analysis of functions on non-compact Lie groups, a mathematical discipline that is still in the early stages of its development. Indeed, many of the facts needed to treat the potential physical applications seem to be still unknown. In this chapter, we will sketch an approach to this theory (inspired, in part, by the recent book by Gel'fand, Graev and Vilenken (1)) that we believe to be most suitable. We shall not treat the functional analysis side of the work very systematically; instead we shall emphasize certain geometric and algebraic insights that do not seem to be adequately treated in the literature. Our main results are concerned with proving a general formula for the characters of induced representations.

In the last section we present a general framework in which one may consider many of these analytic continuation problems.

1. FOURIER TRANSFORM ON GROUPS

The goal of "Fourier analysis on groups" is to represent general classes of complex-valued functions defined on a group space as sums of matrix elements of representations of the group, just as the classical Fourier analysis deals with expansions of functions of one real variable in terms of exponentials. However, there are many technical features of Fourier analysis on non-

THE FOURIER TRANSFORM ON LIE GROUPS

compact, non-abelian groups that are more complicated than for the classical case. Gel'fand, Graev and Vilenkin have given (1) a very clear and detailed introduction to this Fourier analysis on the easiest groups of the semisimple type, SL(2,C) and SL(2,R).

Let us begin by recapitulating what is known about Fourier analysis of functions on a compact group G. We know that the irreducible unitary representation (with two representations identified if they are equivalent) form a discrete set. We can therefore label them by integers, say D_1, D_2,... $g \to D_n(g)$ is a representation of G by unitary operators on a Hilbert space H_n. Suppose $g \to f(g)$ is a continuous function on G. Recall the formula:

$$f(g) = \sum_n \text{tr}(\hat{f}(n)D_n(g))c(n), \quad (1.1)$$

with

$$\hat{f}(n) = \int_G f(g)D_n(g^{-1}) \, dg, \quad (1.2)$$

$$c(n) = \dim H_n.$$

Here, dg is a bi-invariant measure on G, normalized so that the total volume is one.

Let us interpret this in the following way: Regard (1.2) as defining a function of n. It is then an operator-valued function on the "space" of representations of G. Call it the *Fourier transform* of f(g). Then, (1.1) is read as a summation over the representations of G, giving back f(g) in terms of its Fourier transform $\hat{f}(n)$. It is the *inverse Fourier transform*. Thus, it is preferable to consider (1.1-2) as independent formulas, each defining a function of one type in terms of a function of the other type. It is relevant, but not really crucial, that they actually are inverse transformations of each other. Much of our work will be concerned with formal properties of these linear transformations on function spaces, independently of the inverse property.

Of course, the classical Fourier transform of functions of one real variable can be put into the same form. Consider a function f(x), $-\infty < x < \infty$. Define:

$$\hat{f}(y) = \int_{-\infty}^{\infty} f(x) e^{-ixy} \, dx \qquad (1.3)$$

THE FOURIER TRANSFORM ON LIE GROUPS

Here the group G is the additive group of the real numbers. Its irreducible unitary representations are 1-dimensional and parametrized by the real variable y.

$$D_y(x) = e^{ixy}$$

Then,

$$\hat{f}(y) = \int_G f(x) D_y(-x)\, dx \qquad (1.4)$$

(Note that dx is a bi-invariant measure on G.) Here, it is a special feature that all representations occurring are one-dimensional, and that the space of representations can be identified as a space of points with G itself. Note the similarity of (1.3-4) to (1.2). Conversely, to define an analogue of (1.1), given a function $\hat{f}(y)$ define $f(x)$ as follows:

$$f(x) = \frac{1}{2\pi} \int_{-\infty}^{\infty} \hat{f}(y) e^{iyx}\, dy. \qquad (1.5)$$

This is the inverse Fourier transform. Again, these transformations are inverses of each other providing $f(x)$ or $\hat{f}(y)$ satisfy the appropriate conditions

$$\text{(e.g. } \int_{-\infty}^{\infty} (f(x))^2 \, dx).$$

Note that the new feature here is that the space of unitary representations of G is no longer discrete: The summation over this space in (1.1) has been replaced by an integral with respect to an appropriately defined measure.

Now, let us turn to the formulation of these ideas for a general group G (say a Lie group). Let Λ be a space. Assume that to each point $\lambda \in \Lambda$ we have assigned a representation $g \to D_\lambda(g)$ of G by operators on a Hilbert space H_λ. Given a complex-valued function $f(g)$ on G, define its *Fourier transform* $\hat{f}(\lambda)$ by the formula:

$$\hat{f}(\lambda) = \int_G f(g) D_\lambda(g^{-1}) \, dg. \qquad (1.5)$$

Suppose that dg is a right-invariant volume element on G.

Again, $\hat{f}(\lambda)$ is an operator on H_λ. Suppose that $\lambda \to \hat{f}(\lambda)$ is such an operator valued function on Λ. Define the *inverse Fourier transform* as

THE FOURIER TRANSFORM ON LIE GROUPS

follows:

$$f(g) = \int_{\Lambda_0} \text{trace}(\hat{f}(\lambda)D_\lambda(g))c(\lambda)d\lambda. \qquad (1.6)$$

Here, Λ_0 is a subset of Λ, $d\lambda$ is a measure on Λ_0, $c(\lambda)$ is a function on Λ_0. For the moment, we will not assume that (1.5-6) are actually inverses of each other, but shall study them separately as operators.

Notice that (1.5-7) have certain transformation properties. For $g_0 \in G$, define the right-translation, $R_{g_0}(f)$, of f as follows:

$$R_{g_0}(f)(g) = f(gg_0).$$

Then,

$$\widehat{R_{g_0}(f)}(\lambda) = \int_G f(gg_0)D_\lambda(g^{-1})dg$$

$$= \int_G f(gg_0)D_\lambda(g_0(gg_0)^{-1})dg$$

$$= D_\lambda(g_0)\hat{f}(\lambda). \qquad (1.7)$$

For $X \in \underset{\sim}{G}$, define $R_X(f)$ as follows:

$$R_X(f)(g) = \tfrac{\partial}{\partial t} f(g\exp(tX))|_{t=0}.$$

Define:

$$D_\lambda(X) = \frac{\partial}{\partial t} D_\lambda(\exp(tX))\big|_{t=0}.$$

Then,

$$\widehat{R_X(f)}(\lambda) = \int_G R_X(f)(g) D_\lambda(g^{-1}) dg$$

$$= \frac{\partial}{\partial t} \int_G R_{\exp(tX)}(f) D_\lambda(g^{-1}) dg\big|_{t=0}$$

(providing the interchange of integral and derivation is justified, of course)

$$= \frac{\partial}{\partial t} D_\lambda(\exp(tX)) \hat{f}(\lambda)\big|_{t=0}$$

$$= D_\lambda(X) \hat{f}(\lambda) \tag{1.8}$$

This rule is the generalization of the following rule for the classical Fourier transform: The Fourier transform of df/dx is ix times the Fourier transform of f.

So far, we have ignored possible convergence problems concerning the integrals in (1.5) and (1.6), or in the trace of operators. Now we turn to the description of the "generalized function" approach to this point.

THE FOURIER TRANSFORM ON LIE GROUPS

2. THE GENERALIZED-FUNCTION APPROACH TO THE FOURIER TRANSFORM

Suppose that (1.5-6) make sense for a class of functions $g \to f(g)$ and $\lambda \to \hat{f}(\lambda)$ that we will call "test functions." Suppose that $g \to f_1(g)$ and $\lambda \to \hat{f}_1(\lambda)$ are functions that do not necessarily belong to these test-function classes. However, we will suppose that the following "inner products" between these functions and the test-functions do exist:

$$<f_1|f> = \int_G f_1(g)^* f(g) dg \qquad (2.1)$$

$$<\hat{f}_1|\hat{f}> = \int_{\Lambda_0} tr(\hat{f}_1(\lambda)^* \hat{f}(\lambda)) d\lambda. \qquad (2.2)$$

<u>DEFINITION</u>. The generalized function $\lambda \to \hat{f}_1(\lambda)$ is the Fourier transform of the generalized function $g \to f_1(g)$ if:

$$<\hat{f}_1|\hat{f}> = <f_1|f> \qquad (2.3)$$

for every test-function $\lambda \to \hat{f}(\lambda)$, where $g \to f(g)$ is defined in terms of $\hat{f}$ by (1.6).

The generalized function $g \to f_1(g)$ is the

inverse Fourier transform of the generalized function $\lambda \to \hat{f}_1(\lambda)$ if:

$$\langle f_1 | f \rangle = \langle \hat{f}_1 | \hat{f} \rangle,$$

for every test-function $g \to f(g)$ where $\lambda \to \hat{f}(\lambda)$ is defined by (1.5). For example, we can consider the case where G is the additive group of the real line. Then, points of G will be denoted by x, points of Λ by y. $D_y(x)$ is e^{ixy}. Λ_0 is the real y-axis.

Suppose $f_1(x)$ is the Dirac delta function, δ_{x_0}, i.e.

$$\langle f_1(x) | f(x) \rangle = f(x_0)$$

for each test-function $f(x)$. Suppose

$$f(x) = \frac{1}{2\pi} \int_{-\infty}^{\infty} \hat{f}(y) e^{ixy} \, dy. \qquad (2.4)$$

Then, if $\hat{f}_1$ is the Fourier transform of f_1,

$$\langle \hat{f}_1 | \hat{f} \rangle = \langle f_1 | f \rangle = f(x_0) = \frac{1}{2\pi} \int_{-\infty}^{\infty} \hat{f}(y) e^{ix_0 y} \, dy.$$

The right hand side has the form of an inner

THE FOURIER TRANSFORM ON LIE GROUPS

product, and we can read off:

$$\hat{f}(y) = e^{-ix_0 y}.$$

Of course, this agrees with the formal calculation made as if δ_{x_0} were an ordinary function:

$$\hat{\delta}_{x_0}(y) = \int e^{-ixy} \delta_{x_0}(x)\, dx$$

Let us return to a general group G. Consider the case of a function $g \to f(g)$ defined as follows:

$$f'(g) = \operatorname{tr}(FD_\mu(g))c(\mu), \qquad (2.5)$$

where μ is a point of $\Lambda - \Lambda_0$, and F is a given operator on H. Let us suppose that the parameter space Λ is the complex plane. Experience with the classical Fourier transform suggests that the following function should be its Fourier transform:

$$\hat{f}'(\lambda) = \frac{F}{\lambda - \mu} \qquad (2.6)$$

Let us see what is necessary to verify this using the above definition. Now, suppose f(g) is a test-function.

$$\hat{f}(\lambda) = \int_G f(g) D^\lambda(g^{-1}) dg.$$

$$<\hat{f}(\lambda)|\hat{f}'(\lambda)> = \int_{\Lambda_0} tr(\int_G f(g)^* D_\lambda(g^{-1})^* dg \cdot \frac{F}{\lambda - \mu})$$

$$= \int_\Lambda tr(\int_G f(g)^* D_\lambda(g) \frac{F}{\lambda - \mu}) \, dg C(\lambda) d\lambda, \qquad (2.7)$$

supposing that $D_\lambda(g)$ is unitary for $\lambda \in \Lambda_0$. On the other hand,

$$<f|f'> = \int_G f(g)^* tr(F D_\mu(g)) C(\mu) dg \qquad (2.8)$$

In order to have (2.7) equal to (2.8) it is *plausible* that the outer integral in (2.8) be evaluated by "residues," resulting in (2.8). Of course, this requires that $\int_G f(g)^* D_\lambda(g) dg = \hat{f}(\lambda)^*$ fall off sufficiently fast as $\lambda \to \infty$, when $f(g)$ belongs to the chosen class of test functions. This is, in fact, a key property that one requires of the test functions. Gel'fand, Graev and Vilenken have discussed this point for the use $G = SL(2,C)(1)$. The key to their method is to write $\hat{f}(\lambda)$ as an integral operator.

There is another method that can be used, however, at least for those groups having the property that the values of the Casimir operators serve to characterize the representation (e.g., the semisimple group). Let Δ be a Casimir operator of $\underset{\sim}{G}$, i.e. an element of $U(\underset{\sim}{G})$ (= the universal enveloping algebra of $\underset{\sim}{G}$) that commutes with all elements of $\underset{\sim}{G}$. Suppose that $D_\lambda(\Delta)$ is equal to $\Delta(\lambda)$, where $\lambda \to \Delta(\lambda)$ is a complex-valued scalar function on Λ. Then, we have:

$$\widehat{L_\Delta(f)}(\lambda) = \pm \hat{f}(\lambda)\Delta(\lambda),$$

or

$$\hat{f}(\lambda) = \frac{\widehat{L_\Delta(f)}(\lambda)}{\Delta(\lambda)}$$

Thus, if $g \to f(g)$ is rapidly decreasing on G in the sense that the Fourier transform of it and its derivatives exist and are bounded, and if $\Delta(\lambda) \to \infty$ as $\lambda \to \infty$, then $\hat{f}(\lambda) \to 0$ as $\lambda \to \infty$.

One of the main problems in this approach to Fourier analysis on groups is to determine precisely which classes of test-functions correspond under Fourier and inverse-Fourier transform.

3. INTEGRALS ON HOMOGENEOUS SPACES

Before proceeding further with the study of the Fourier transform, we must present some material concerning integrals on groups and homogeneous spaces. Suppose a Lie group G acts transitively on a space M. Let p_0 be a fixed point of M, and let L be the isotropy subgroup of G at p_0. Then, M can be considered as the coset space G/L.

Suppose dg is a bi-invariant volume element on G, and that dp is a volume element on M. Let π be the map $G \to M$ defined by

$$\pi(g) = gp_0.$$

Let L_g and R_g denote the left and right translation on G, i.e.

$$L_g(g') = gg' \quad \text{for} \quad g, g' \in G$$
$$R_g(g') = g'g^{-1} \quad \text{for} \quad g, g' \in G.$$

Let T_g denote translation by g on M, i.e.

$$T_g(p) = gp \quad \text{for} \quad p \in M.$$

THE FOURIER TRANSFORM ON LIE GROUPS 65

Let $J_g(p)$ be the Jacobian of T_g with respect to the volume element dp, i.e.

$$\int_M f(p)dp = \int_M f(T_g p) J_g(p) dp \qquad (3.1)$$

for each function $p \to f(p)$ on M. The following rule holds

$$J_{g_1 g_2}(p) = J_{g_2}(p) J_{g_1}(g_2 p) \qquad (3.2)$$

for g_1, $g_2 \in G$, $p \in M$.

Notice also that:

$$\pi(L_g(g')) = T_g \pi(g') \quad \text{for} \quad g, g' \in . \qquad (3.3)$$

Our aim is to express the integral $\int_G f(g)dg$ of a function on G as an iterated integral over L and over M. We will do this with a special assumption (that will be satisfied in the cases we have in mind), namely, suppose there is a subset M' of M such that

> M - M' has measure zero with respect to the volume element dp. (3.4)

There is a cross-section map $\phi M' \to G$,

such that $\pi\phi(p) = p$ for $p \in M'$,
and $\phi(p_0) = e$. (3.5)

Given a function $f(g)$ on G, let us try to express $\int_G f(g)dg$ via the following Ansatz:

$$\int_G f(g)dg = \int_{M'} (\int_L f(\phi(p)\ell)h(\phi(p)\ell)d\ell)dp. \quad (3.6)$$

Here, $d\ell$ is the left-invariant volume element on L, $h(g)$ is some function on G, such that (3.6) is to hold for all functions $g \to f(g)$.

Our aim is to calculate $h(g)$ by working out the conditions imposed by the bi-invariance of dg and the left invariance of $d\ell$. However, we will not assume that $d\ell$ is right invariant in L. Let us work out first the condition this imposes.

Suppose ℓ_0 is a fixed element of L, and $f(\ell)$ is a function on L. Define a volume element $d\ell'$ on L by the following rules:

$$\int_L f(\ell)d\ell' = \int_L f(\ell\ell_0)d\ell.$$

Note that it is still a left-invariant volume element on L, hence there is a real *constant* $a(\ell_0)$

THE FOURIER TRANSFORM ON LIE GROUPS

such that

$$d\ell' = a(\ell_0)d\ell, \quad \text{i.e.}$$

$$\int_L f(\ell\ell_0)d\ell = a(\ell_0) \int_L f(\ell)d\ell. \qquad (3.7)$$

(Stated another way, $a(\ell_0)$ is the Jacobian with respect to $d\ell$ of the *right* translation by ℓ.) One can show that

$$a(\ell_0\ell_1) = a(\ell_0)a(\ell_1), \qquad (3.8)$$

i.e. $\ell \to a(\ell)$ is a homomorphism of L into the real numbers. Now, using (3.6),

$$\int_G f(g\ell_0)dg = \int_{M'} (\int_L f(\phi(p)\ell\ell_0)h(\phi(p)\ell)d\ell dp$$

$$= a(\ell_0) \int_{M'} \int_L f(\phi(p)\ell)h(\phi(p)\ell\ell_0^{-1})d\ell dp,$$

hence, right invariance of dg forces

$$a(\ell_0)h(\phi(p)\ell\ell_0^{-1}) = h(\phi(p)\ell).$$

Set $\ell = e$ in this relation, giving

$$h(\phi(p)\ell) = h(\phi(p))a(\ell^{-1}) \quad \text{for} \quad \ell \in L. \qquad (3.9)$$

Using this in (3.6) gives

$$\int_G f(g)dg = \int_{M'} \int_L f(\phi(p)\ell)h(\phi(p))a(\ell^{-1})d\ell dp. \qquad (3.10)$$

Now, let us use left-invariance of dg. For $g_0 \in G$,

$$\int_G f(g_0 g)dg = \int_{M'} \int_L f(g_0\phi(p)\ell)h(\phi(p))a(\ell^{-1})d\ell dp.$$

Now,

$$\pi(g_0\phi(p)) = g_0 \pi\phi(p) = g_0 p = \pi\phi(g_0 p), \text{ here}$$

or

$$g_0\phi(p) = \phi(g_0 p)\ell(p), \qquad (3.11)$$

where $p \to \ell(p)$ is a function from M' to L. Thus,

$$\int_G f(g_0 g)dg = \int_{M'} \int_L f(\phi(g_0 p)\ell(p)\ell)h(\phi(p))$$
$$\qquad a(\ell^{-1})d\ell dp$$
$$= \int_{M'} \int_L f(\phi(g_0 p))h(\phi(p))a(\ell^{-1}\ell(p))d\ell dp$$
$$= \int_{M'} \int_L f(\phi(g_0 p)\ell)h(\phi(p))a(\ell^{-1})a(\ell(p))d\ell dp$$
$$= \int_{M'} \int_L f(\phi(p)\ell)h(\phi(g_0^{-1}p))a(\ell_{-1})$$
$$\qquad a(\ell(g_0^{-1}p))J_{g_0^{-1}}(p)d\ell dp$$

THE FOURIER TRANSFORM ON LIE GROUPS 69

Since f is to be an arbitrary function, this forces

$$h(\phi(g_0^{-1}p))a(\ell^{-1})a(\ell(g_0^{-1}p))J_{g_0^{-1}}(p)$$

$$= h(\phi(p))a(\ell^{-1}), \text{ or}$$

$$h(\phi(g_0^{-1}p))a(\ell(g_0^{-1}p))J_{g_0^{-1}}(p) = h(\phi(p)) \quad (3.12)$$

Set p_0 in this relation. Now, we may normalize things so that

$$\phi(p_0) = e,$$
$$h(e) = 1. \quad (3.13)$$

Using (3.13) in (3.12) gives

$$1 = J_{g_0^{-1}}(p_0)h(\phi(g_0^{-1}p_0)a(\ell(g_0^{-1}p_0))^{-1}) \quad (3.14)$$

Using (3.11) gives

$$\ell(g_0^{-1}p_0) = g_0\phi(g_0^{-1}p_0)$$

Using this in (3.14) gives

$$1 = J_{g_0^{-1}}(p_0)h(g_0^{-1}),$$

hence we have the final result

$$h(g) = J_g(p_0)^{-1} \qquad (3.15)$$

and

$$\int_G f(g)dg = \int_{M'} \int_L f(\phi(p)\ell) J_{\phi(p)}(p_0)^{-1}$$

$$a(\ell^{-1}) d\ell dp \qquad (3.16)$$

Of course, this has been derived on the assumption that such a formula exists. One can, in fact prove that it does, using the theory of integration of differential forms on manifolds.

4. THE FOURIER TRANSFORM AS AN INTEGRAL OPERATOR

Suppose that a Lie group G acts as a transformation groups on a space M. Let $F(M)$ be a space of complex-valued functions on M. Define a one-parameter family $g \to D_\lambda(g)$ of representations of G by operators in $F(M)$ as follows:

$$D_\lambda(g)(\Psi)(p) = m(g, p)^\lambda \Psi(g^{-1}p) \qquad (4.1)$$

λ is a complex number; Ψ a function in $F(M)$; p

THE FOURIER TRANSFORM ON LIE GROUPS

a point of M. $m(g, p)$ is a multiplier system (see (4)), i.e.

$$m(g_0 g_1, p) = m(g_0, p) m(g_1, g_0^{-1} p)$$

$$\text{for} \quad g_0, g_1 \in G, p \in M. \tag{4.2}$$

If $g \to f(g)$ is a function on G, define its Fourier transform $\hat{f}(\lambda)$ as an operator: $F(M) \to F(M)$

$$\hat{f}(\lambda)(\Psi) = \int_G f(g) D_\lambda(g^{-1})(\Psi) dg. \tag{4.3}$$

Hence,

$$\hat{f}(\lambda)(\Psi)(p) = \int_G f(g) m(g^{-1}, p)^\lambda \Psi(gp) dg. \tag{4.4}$$

Now, we want to express this in the form:

$$\hat{f}(\lambda)(\Psi)(p) = \int_{M'} k(p, q; \lambda, f) \Psi(q) dq, \tag{4.5}$$

where the kernel $k(p, q; \lambda, f)$, of this integral operator is defined by f, and where dq is a given volume element on M. We shall suppose that G acts transitively on M.

Introduce $\pi = G \to M$, M', $\phi: M' \to G$ as in Section 3. Then, for $p \in M'$,

$$\hat{f}(\lambda)(\Psi)(p) = \int_G f(g)m(g^{-1}, p)^\lambda \Psi(g\pi\phi(p))dg$$

$$= \int_G f(g)m(g^{-1}, p)^\lambda \Psi(\pi g\phi(p))dg$$

$$= \int_G f(g\phi(p)^{-1})m(\phi(p)g^{-1}, p)^\lambda \Psi(\pi g)dg$$

$$= \text{using (3.16)},$$

$$\int_{M'} \int_L f(\phi(q)\ell\phi(p)^{-1})m(\phi(p)\ell^{-1}\phi(q), p)^\lambda$$

$$\Psi(\pi(\phi(q)\ell))$$

$$J_{\phi(q)}(p_0)^{-1} a(\ell^{-1}) d\ell dq$$

Now, $\pi(\phi(q)\ell) = \pi(\phi(q)) = q$. Then, we see that (4.5) holds, with

$$k(p, q; \lambda, f) = \int_L f(\phi(q)\ell\phi(p)^{-1})$$

$$m(\phi(p)\ell^{-1}\phi(q), p)^\lambda \qquad (4.6)$$

$$J_{\phi(q)}(p_0)^{-1} a(\ell^{-1}) d\ell$$

Thus, we see that it follows from rather general principles that the Fourier transform of groups whose representations as given by induced-representation-homogeneous vector bundle-theory can be described by integral operators. In this section, we have considered, for simplicity of

THE FOURIER TRANSFORM ON LIE GROUPS   73

notation, only trivial, direct product, vector bundles, although the ideas carry over to the general case.

5. CHARACTERS OF MULTIPLIER REPRESENTATIONS

Continue with the notations of the previous section. We will now show that formula (4.6) provides us with a convenient way to calculate the character of the representation D_λ defined by (4.1). First, however, we will recall the general principles of character theory for representations of non-compact groups.

Suppose D: $g \to D(g)$ is a representation of a group G on a vector space H. If $g \to f(g)$ is a function on G, define:

$$D(\hat{f}) = \int_G f(g)D(g)dg. \qquad (5.1)$$

Let $\chi(\hat{f})$ be the "trace" of this operator. Rather than get involved with technicalities of functional analysis we will consider a special case adequate for the applications we have in mind, namely that where H is the space of complex-valued functions on a space M', and each $\hat{f}$ is realized as an integral operator on H with respect to a measure dp on M:

$$\hat{f}(\Psi)(p) = \int_{M'} k(p, q; f)\Psi(q)dq.$$

Then,

$$\chi(\hat{f}) = \int_{M'} k(p, p; f)dp. \qquad (5.2)$$

Now, $\hat{f} \to \chi(f)$ will be a linear functional on the space of functions f that we are considering. The *character* of D, denoted by $\chi(g)$, will be a function on G such that:

$$\chi(f) = \int_G f(g)\chi(g)dg \qquad (5.3)$$

Let us now assume that the kernel k is given by (4.6). Then,

$$\chi(f) = \int_{M'} \int_L f(\phi(p)\ell\phi(p)^{-1})m(\phi(p)\ell^{-1}\phi(p), p)^\lambda$$

$$J_{\phi(p)}(p_0)^{-1}a(\ell^{-1})d\ell dp \qquad (5.4)$$

We now hope to compare (5.4) with (5.3) to calculate $\chi(g)$.

Consider the map Φ: M' × L → G defined as

THE FOURIER TRANSFORM ON LIE GROUPS

follows:

$$\Phi(p, \ell) = \phi(p)\ell\phi(p)^{-1}$$

for $p \in M'$, $\ell \in L$.

Then, there is a function $\alpha(p, \ell)$ on $M' \times L$ such that

$$\int_{M' \times L} f(p, \ell)\,dp\,d\ell = \int_G \left(\sum_{(p,\ell) \in \Phi^{-1}(g)} \frac{f(p,\ell)}{\alpha(p,\ell)} \right) dg \quad (5.5)$$

for every function f on $M' \times L$. α is the factor of proportionality between the volume elements $dp\,d\ell$ and the volume element dg pulled back via Φ. (The sum is necessary because Φ is not necessarily one-one.)

Let us apply (5.5) to (5.4):

$$\chi(f) = \int_G \left(\sum_{(p,\ell) \in \Phi^{-1}(g)} f(g)m(g, p)^\lambda \right.$$
$$\left. J_{\phi(p)}(p_0)^{-1} a(\ell^{-1}) \alpha(p, \ell)^{-1} \right) dg$$

$$= \int_G f(g) \left(\sum_{(p,\ell) \in \Phi^{-1}(g)} m(g, p)^\lambda \right.$$
$$\left. J_{\phi(p)}(p_0)^{-1} a(\ell^{-1}) \alpha(p, \ell)^{-1} \right) dg.$$

Comparing with (5.3) shows that:

$$\chi(g) = \sum_{(p,\ell)\varepsilon\Phi^{-1}(g)} m(g, p)^\lambda$$

$$J_{\phi(p)}(p_0)^{-1} a(\ell^{-1}) \alpha(p, \ell)^{-1} \qquad (5.6)$$

This is the formula for the character. (There are many special cases in (2), and there one will find detailed calculations of the various terms for the classical groups.)

6. RELATIONS BETWEEN THE FOURIER EXPANSIONS OF FUNCTIONS ON DIFFERENT REAL FORMS OF THE SAME COMPLEX GROUP

Let G be a group, G_c its complexification, G_μ a compact subgroup of G_c that is also a real form of G_c. Suppose there is a Cauchy-like integral formula converting a function $f(g_0)$ on G_c into a complex-analytic function $f(g_c)$ on $G - G_0$:

$$f(g_c) = \int_G C(g_c, g) f(g) dg \qquad (6.1)$$

THE FOURIER TRANSFORM ON LIE GROUPS

Here, dg is the right-invariant volume element on G.

Let us suppose that the integral formula (6.1) is equivariant with respect to right-translation by elements of G:

$$f(g_c g_0) = \int_G C(g_c g_0, g) f(g) dg$$

$$= \int_G C(g_c, g) f(g g_0) dg$$

$$= \int_G C(g_c, g g_0^{-1}) f(g) dg, \text{ i.e.}$$

$$C(g_c g_0, g) = C(g_c, g g_0^{-1}) \text{ for}$$

$$g_c \in G_c, g, g_0 \in G. \qquad (6.2)$$

Let us now formulate the Fourier expansion of $f(g)$ in terms of the representations of G. Suppose that Λ is a parameter space. For each $\lambda \in \Lambda$, suppose given a Hilbert space H_λ and a representation $g \to D_\lambda(g)$ of G by operators on H_λ. Given a function $f(g)$ on G, define

$$\hat{f}(\lambda) = \int_G f(g) D_\lambda(g^{-1}) dg$$

Conversely, given an operator-valued function

$\lambda \to \hat{f}(\lambda)$ on Λ, define:

$$f(g) = \int_{\Lambda_0} tr(\hat{f}(\lambda)D_\lambda(g))c(\lambda)d\lambda, \qquad (6.3)$$

where $d\lambda$ is a volume-element on Λ_0, a subset of Λ, $\lambda \to c(\lambda)$ is a function on Λ, possibly having singularities off Λ_0. Substitute (6.3) in (6.1) and interchange integrals:

$$f(g_c) = \int_{\Lambda_0} tr(\hat{f}(\lambda)\int_G C(g_c, g)D_\lambda(g)dg)c(\lambda)d\lambda.$$

This suggests that we set:

$$D_\lambda(g_c) = \int_G C(g_c, g)D_\lambda(g)dg. \qquad (6.4)$$

Then,

$$f(g_c) = \int_{\Lambda_0} tr(\hat{f}(\lambda)D_\lambda(g_c))c(\lambda)d\lambda. \qquad (6.5)$$

Suppose now that, in this formula, we consider $g_c = g_\mu \in G_\mu$. We expect that (6.5) should generate the Fourier expansion of $f(g_\mu)$ relative to the irreducible representations of G_μ, which of course form a discrete set. Suppose, in fact, that $\lambda_1, \lambda_2, \ldots$ are the values of λ corresponding to

THE FOURIER TRANSFORM ON LIE GROUPS 79

the finite dimensional representations of G, which are also the values corresponding to the finite dimensional unitary representations of G_μ.

We can now suggest the following program: Show that the operator-valued function $\lambda \to D_\lambda(g_\mu)c(\lambda)$ is meromorphic (at least as a generalized-function, i.e. when it occurs in an integral of the form (6.5)), and show that the poles occur at the values $\lambda = \lambda_1, \lambda_2, \ldots$. Further, the residues at these poles should be the operators corresponding to the finite dimensional representations of G_μ.

Notice that there will be a crucial distinction between the cases where the "Plancherel measure" $c(\lambda)$ is smooth on all of Λ, or develops singularities off of the subset Λ_0. In fact, for $G = SL(2,C)$, Λ is a space of two complex variables, and $c(\lambda)$ is indeed complex-analytic over all of Λ, and takes as its value at $\lambda = \lambda_n$ the dimension of the corresponding irreducible representation of $G_\mu = SU(2) \times SU(2)$. In fact has been noted by Gel'fand, Graev and Vilenkin (1), but it plays no special role in their work. On the other hand, for $G = SO(2,1)$, Λ is the complex plane: $c(\lambda)$ is meromorphic on Λ, and its residues at the λ_n are the dimensions of the corresponding irreducible representations of G_μ = SO(3,R). The distinction between these two cases also plays a role in the work of Toller on S-matrix theory.

That these ideas are at least qualitatively appropriate can be seen by referring to (1), Chapter 15. The relevant group is SO(3,R) = G_μ and SO(2,1) = G. In fact, $P_n(z)$ are the matrix-elements of representations of SO(3,R), for $-1 < z < 1$. They can be extended analytically in n and z, with a discontinuity across the part of the z-plane corresponding to G, i.e. $1 \leqq z < \infty$. Note also that the Legendre functions of the second kind $Q_n(z)$ have as discontinuities across the line $-1 \leqq z \leqq 1$ the $P_n(z)$: This indicates that one might also use a Cauchy integral formula involving integration over G_μ instead of over G. These ideas will be explored further in a later chapter.

7. PROPERTIES OF THE CAUCHY KERNEL

Keep the notations of the last section. Suppose that $g \to D(g)$ is a representation of the

THE FOURIER TRANSFORM ON LIE GROUPS 81

non-compact real-form group G. Let $C(g_c, g)$ be a Cauchy-type kernel with the properties described in the previous section. From

$$D_1(g_c) = \int_G C(g_c, g)D(g)dg \qquad (7.1)$$

We would like to find what conditions C must satisfy in order that $D_1(g_c)$ satisfy

$$D_1(g_c g_c') = D_1(g_c)D_1(g_c') \qquad (7.2)$$

whenever $g_c, g_c' \in G_c - G$ are such that $g_c g_c' \in G_c - G$. (We are also regarding (7.1) as a "formal" expression, i.e. we are ignoring convergence questions. Even if (7.1) should not define the left-hand side as a genuine operator-valued function of g_c, one can usually make some sense of these relations using the generalized-function idea.) Then,

$$D_1(g_c)D_1(g_c') = \int_{G \times G} \int C(g_c, g)$$
$$C(g_c', g')D(g)D(g')dgdg'$$
$$= \iint C(g_c, g)C(g_c', g')D(gg')dgdg'$$
$$= \iint C(g_c, g)C(g_c', g^{-1}g')D(g')dgdg'$$

If (7.2) is satisfied, then the left-hand side also equals

$$\int C(g_c g_c', g') D(g') dg'$$

(7.2) will then be satisfied (at least formally) if:

$$C(g_c g_c', g') = \int_G C(g_c, g) C(g_c', g^{-1} g') dg \qquad (7.3)$$

This can be interpreted in terms of the convolution product of functions on G. Suppose that $f(g)$, $h(g)$ are functions on G. Define $f * h$, the *convolution* of f and h, as the following function on G:

$$f * h(g') = \int_G f(g) h(g^{-1} g') dg'. \qquad (7.4)$$

Then, for each g_c, define C_{g_c} as the following function on G'.

$$C_{g_c}(g') = C(g_c, g')$$

We see that the condition for (7.2) is that the mapping $g_c \to C_{g_c}$ be a "homomorphism" of a subset of $G_c - G$ into the convolution-product algebra of

THE FOURIER TRANSFORM ON LIE GROUPS 83

functions on G. (Of course, to make rigorous sense of this, it will be necessary to regard $C(g_c, g)$ as a generalized function on $G_c - G) \times G$, i.e. as a linear functional on a certain class of functions on $(G_c - G) \times G$.)

EXAMPLE. <u>The additive group of the complex numbers</u>.

This example is rather trivial, since the group is abelian, but will serve to illustrate the ideas.

G_c consists of the complex numbers z, G of the real numbers, a typical one denoted by t. G_μ now should be interpreted as the pure imaginary numbers. The invariant measure on G is dt, and an obvious guess for the Cauchy kernel is:

$$C(z, t) = \frac{1}{(2\pi i)(t - z)}$$

Let us check (7.3):

$$\int_G C(g_c, g) C(g_c', g^{-1}g') dg$$

$$= \frac{1}{(2\pi i)^2} \int_{-\infty}^{\infty} \frac{dt}{(t - z)(t' - t - z')} \qquad (7.5)$$

Suppose, for example, that z' is in the upper half-plane. Then, (7.5) can be evaluated by finding the residues of the function

$$t \to \frac{1}{(t - z)(t - z' + t')}$$

in the upper half-plane. (Since the contour integral of the integrand over a semicircle in the complex-t plane goes to zero as the radius of the semicircle goes to infinity.)

Case 1. z also in upper half-plane.

The pole is at t = z. Hence, (7.5) equals

$$\frac{1}{2\pi i} \frac{1}{(t' - z - z')} = C(z + z', t').$$

Case 2. z in the lower half plane.

There are no poles in the upper half-plane: (7.5) is zero. Hence, (7.3) is only true if g_c and g_c' both belong to a certain region of G_c (the upper half-plane, in this case). Notice that this region forms a semi-subgroup of G_c. Thus we conjecture, in general, that (7.3) will hold when g_c and g_c' belong to certain sub-semigroups of the

THE FOURIER TRANSFORM ON LIE GROUPS 85

group G_c. (A sub-semigroup is a subset of a group that is closed under multiplication, but not under inverses.)

Let us now calculate

$$D_\lambda(g_c) = \int_G C(g_c, g) D_\lambda(g) dg.$$

In this case, λ is a complex number, and, with the identification $g = t$, $g_c = z$,

$$D_\lambda(t) = e^{\lambda t},$$

$$D_\lambda(z) = \frac{1}{2\pi i} \int_{-\infty}^{\infty} \frac{e^{\lambda t}}{t - z} dt = e^{\lambda z}$$

BIBLIOGRAPHY

1. I. M. Gel'fand, M. I. Graev, and N. Ya. Vilenkin, Generalized Functions, Vol. 5, Academic Press, New York, 1966.
2. I. M. Gel'fand and M. A. Naimark, Unitare Darstellungen der Klassichen Gruppen, Akademie-Verlag, Berlin, 1957.
3. I. M. Gel'fand and G. E. Shilov, Generalized Functions, Vol. 1, Academic Press, New York, 1964.
4. R. Hermann, Lie Groups for Physicists, W. A. Benjamin, New York, 1966.
5. _____, Analytic Continuation of Group Representations, part V, Communications in Math. Phys. $\underline{5}$, 157 (1967).
6. E. T. Whittaker and G. N. Watson, A Course of Modern Analysis, Cambridge, 1963.

CHAPTER III

CAUCHY INTEGRALS ON LIE GROUPS AND MATRIX ELEMENT FUNCTIONS OF THE SECOND KIND

1. INTRODUCTION

In this book we cover two interrelated topics: The theory of Cauchy integral kernels for Lie groups (continuing work of Chapter II) and the theory of "matrix-element functions of the second kind" [1]. Our main aim is to interpret group-theoretically, and thereby generalize, two classical formulas concerning special functions:

$$Q_\ell(z) = \int_{-1}^{1} \frac{P_\ell(\mu)}{z - \mu} \, d\mu$$

$$\frac{1}{z - \mu} = \sum_{\ell} Q_\ell(z) P_\ell(\mu)(2\ell + 1) \qquad (1.1)$$

Here, P_ℓ and Q_ℓ are Legendre functions of the first and second kind.

Now it is well-known how the $P_\ell(\mu)$ are related to matrix elements of representations of $SO(3,R)$, enabling one to derive most of the important properties of these functions using group-theoretical reasoning. A similar interpretation of the Q_ℓ as matrix elements does not seem to be in the literature; we will provide such an interpretation in this book. In fact, we will show that the Q's can be described as a special case of the following general situation. Let G be a group, with H a complex vector space. Suppose that $D(G)$ is a representation of G by operators on H, and that a Hermitian bilinear form $\langle \Psi | \Psi' \rangle$, $\Psi, \Psi' \in H$, is defined on H. (However, we do not necessarily require that this form be positive definite, nor even that its values be finite numbers, i.e. we allow ∞ as a possible value.) Suppose that, for each $g \in G$, there is a unique adjoint operator $D(g)^*$ such that:

$$\langle D(g)^* \Psi | \Psi' \rangle = \langle \Psi | D(g) \Psi' \rangle \quad \text{for}$$
$$\Psi, \Psi' \in H.$$

Then, $g \to D'(g) = D(g^{-1})^*$ defines another representation of G by operators on H.

Suppose now that H' is a finite-dimensional subspace of H that is left invariant by D'(G). Suppose (Ψ_j), $1 \leq j \leq n$, is a basis for H'. Set:

$$e_{jk}(g) = \langle \Psi_j | D(g) \Psi_k \rangle.$$

Now, we do not suppose that D(G) leaves H' invariant, hence $g \to (e_{jk}(g))$ is not necessarily a matrix representation of G. However, those functions transform by a finite dimensional representation of G under left-translation by G (but not right translation). Suppose, in fact, that:

$$D'(g)(\Psi_j) = \sum_k d_{kj}(g) \Psi_k.$$

Then, $g \to (d_{jk}(g))$ is indeed a matrix representation of G. Further, for $g_0 \in G$,

$$e_{jk}(g_0^{-1} g) = \langle \Psi_j | D(g_0^{-1}) D(g) \Psi_k \rangle$$

$$= \langle D(g_0^{-1})^* \Psi_j | D(g)\Psi_k \rangle$$

$$= \langle D'(g_0)\Psi_j | D(g)\Psi_k \rangle$$

$$= \sum_{k'} d_{k'j}(g_0)^* \langle \Psi_{k'} | D(g)\Psi_k \rangle$$

$$= \sum_{k'} d_{k'j}(g_0)^* e_{k'k}(g)$$

(* applied to a complex number denotes complex-conjugation)

We see indeed that the functions $e_{k'k}(g)$ transform under left-translation by a representation of G.

If, further, K is a subgroup of G such that $D(K)(H') = H'$, we see, in a similar way that the e's transform under right-translation by elements of K via a matrix representation of K.

Thus, we see that whenever we have a representation $g \to D(g)$ of G such that $D'(G)$ is reducible, this method applies to define "matrix-element functions of the second kind." Of course, if G is semisimple and H is finite dimensional, these functions are not interesting, since the representation is completely reducible. The discussion given in [3, 4] provides methods for constructing representations of many groups with these properties.

I would like to thank C. Itzykson, E. Stein and E. Wichmann for their helpful suggestions concerning this material. The paper by Stein and Wainger [6] has been very helpful in my work, and many of my ideas are reformulations of material to be found there in a different language.

2. GENERAL REMARKS CONCERNING CAUCHY KERNELS FOR COMPLEX LIE GROUPS

Let G_c be a complex Lie group, and let G be a real subgroup. Consider a Cauchy-type kernel $C(g_c, g)$ defined for $g_c \in G-S$, $g \in S$, where S is a closed subset of G_c containing G. In addition we suppose that, for fixed g, $g_c \to C(g_c, g)$ is holomorphic.

We can now use C to set up an integral transform:

$$F(g_c) = \int_G C(g_c, g) f(g) dg, \qquad (2.1)$$

converting a continuous function $g \to f(g)$ on G into a function $g_c \to F(g_c)$ holomorphic on $G_c - S$. dg is a bi-invariant volume element on G.

Let us investigate the conditions that this transform intertwine left-translation on G_c and G: Given $g_0 \in G$,

$$F(g_0 g_c) = \int_G C(g_0 g_c, g) f(g) dg$$

$$\int_G C(g_c, g) f(g_0 g) dg$$

$$\int_G C(g_c, g_0^{-1} g) f(g) dg.$$

The intertwining condition is then:

$$C(g_0 g_c, g) = C(g_c, g_0^{-1} g) \qquad (2.2)$$

We will suppose that this condition is satisfied by requiring that $C(g_c, g)$ be of the following form:

$$C(g_c, g) = C(g^{-1} g_c), \qquad (2.3)$$

where $g_c \to C(g_c)$ is a function defined and holomorphic on $G_c - S$. In addition, we will suppose that the singularity set S is mapped into itself by left-translation under G.

Suppose now that K is a subgroup of G. We

CAUCHY INTEGRALS ON LIE GROUPS   93

will examine the conditions that the transform defined by (2.1) also be equivariant with respect to right translation via K:

$$F(g_c k) = \int_G C(g^{-1} g_c k) f(g) dg$$

$$\int_G C(g^{-1} g_c) f(gk) dg = \int_G C((gk^{-1})^{-1} g_c) f(g) dg$$

$$= \int_G C(kg^{-1} g_c) f(g) dg.$$

Hence, the equivariance condition is:

$$C(kg_c k^{-1}) = C(g_c) \quad \text{for} \quad g_c \in G_c, \ k \in K. \tag{2.4}$$

For example, let us consider the case:

$$G = SO(3,R), \ G_c = SO(3,C), \ K = SO(2,R).$$

A *spherical function* of G is a function f(g) that is invariant under left and right translation by K and that is an eigenfunction of the Casimir operator of $\underset{\sim}{G}$, considered as a second-order differential operator acting on the functions on G. Suppose that (2.3) and (2.4) are satisfied. Then, (2.1) converts f into a function $F(g_c)$ that is also an eigenfunction of the Casimir operator (since the transform commutes with left-translation) and that is invariant under left and right translation by K.

Now, describe points of G_c by complex Euler angles (ϕ, θ, Ψ), where ϕ and Ψ are parameters describing left and right translation by K_c (= the complexification of K). Then, $F(g_c)$ and $f(g)$ only depend on θ, and, since both are eigenfunctions of the Casimir operator, both satisfy the same second order ordinary differential equation. However, they are independent solutions, since $F(g_c)$ has a singularity at g_c = e, while $f(g)$ is regular at g = e. We see that F should be the "Legendre function of the second kind," and (2.1) should reduce to (1.1) when z = cos θ, and the integral over G is expressed in terms of Euler angles. (Note, in fact, that d_g = dϕdΨd(cos θ).)

3. HUA'S CONSTRUCTION OF CAUCHY KERNELS

To construct specific examples of Cauchy kernels and related semigroups, we can utilize Hua's work [5]. (However, his construction does

CAUCHY INTEGRALS ON LIE GROUPS

not give the Cauchy kernel for $G_c = SO(3,C)$.)

Suppose that V is a finite dimensional complex vector space. Let E be the vector space of linear operators on V. Suppose also that V has a nondegenerate anti-linear complex-valued bilinear form $v_1, v_2 \to (v_1/v_2)$, i.e.

$$(cv_1/v_2) = c^*(v_1/v_2), \quad (v_1/cv_2) = c(v_1/v_2)$$

for $v_1, v_2 \in V$, complex scalars c. (c^* denotes the complex-conjugate of a complex number.) If $A \in E$, define $A^* \in E$ so that:

$$(v_1/Av_2) = (A^*v_1/v_2) \quad \text{for} \quad v_1, v_2 \in V.$$

(Of course, A^* is just the adjoint of A with respect to the form (/).) Then,

$$(A + B)^* = A^* + B^*$$

$$(cA)^* = c^*A^*$$

$$(AB)^* = B^*A^*$$

for $A, B \in E$, $c \in C$.

Let us say that $A \in E$ is *positive* with respect to the form (/) if:

$(v/Av) > 0$ for all $v \in V$, $v \neq 0$.

Let us denote this by $A > 0$. If $-A > 0$, write $A < 0$. A is unitary with respect to the form if:

$$A^{-1} = A^*, \quad \text{or} \quad {}^*AA - 1 = 0.$$

Define E^+ and E^- as the set of $A \in E$ such that, respectively, such that

$$A^*A - 1 < 0$$
$$A^*A - 1 > 0.$$

THEOREM 3.1. E^+ and E^- are semigroups with respect to operator multiplication.

Proof. Suppose $A, B \in E^{\pm}$. Set:

$$C = A^*A - 1$$

Then,

$$(AB)^*(AB) - 1 = B^*A^*AB - 1$$
$$= B^*(C+1)B - 1$$
$$= B^*CB + B^*B - 1$$

CAUCHY INTEGRALS ON LIE GROUPS 97

If $C > 0$, so is B^*CB. This proves that $AB \in E^{\pm}$.

Now, suppose that G_c is a complex Lie group, and that $\rho: G_c \to E$ is a complex-analytic representation of G_c by operators on V. Suppose G is a subgroup of G_c such that the operators of $D(G)$ are unitary. For g_c, $g \in G_c$, set:

$$C(g_c, g) = (\det(1-\rho(g_c)\rho(g^{-1})))^{-n} \qquad (3.1)$$

Then, following Hua [6], C is a good candidate for a Cauchy kernel; assigning a function $F(g_c)$ to a function $f(g)$ on G:

$$F(g_c) = \int_G C(g_c, g) f(g) dg$$

where dg is invariant measure on G.

This construction applies naturally to the case where $G_c = GL(n, c)$, the group of $n \times n$ complex matrices, with ρ the defining representation of G_c. G can be taken as $U(p, q)$, $p+q = n$, the subgroup leaving invariant a Hermitian form with p minus signs, and q plus signs. However, this construction is not too well adopted to the case $G_c = SL(2, C)$, for example. Rather than try to modify Hua's construction, we will analyze further the conditions that we want the Cauchy kernel to satisfy.

4. EXPANSIONS OF CAUCHY KERNELS OVER COMPACT GROUPS AND WIGNER d-FUNCTIONS OF THE SECOND KIND

Let G be a compact semisimple group, K a subgroup, with G_c its complexification. Suppose $g_c \to C(g_c)$ is a function on an open subset of G_c such that:

a) C is holomorphic.
b) $C(kg_c k^{-1}) = C(g_c)$, for $k \in K$, $g_c \in G_c$.
c) For each g_c in the domain of definition of C, the function $g \to C(g_c g^{-1})$ is infinitely differentiable on G.

Suppose D_j, $j = 0, 1, 2,\ldots$ are the irreducible representations of G by operators on Hilbert spaces H_j. We then have an expansion of the form

$$C(g^{-1}g_c) = \sum_j \text{trace}(E_j(g_c)D_j(g^{-1}))c(j), \quad (4.1)$$

with

CAUCHY INTEGRALS ON LIE GROUPS

$$c(j) = \text{dimension } H_j \qquad (4.2)$$

$$E_j(g_c) = \int_G C(g^{-1}g_c) D_j(g) dg. \qquad (4.3)$$

(Here, dg is the bi-invariant volume element on G, normalized so that the total volume of G is one.)

The functions $E_j(g_c)$ may be thought of as analogous to the Legendre functions of the second kind. Notice that the relations (4.1)-(4.3) imply an interdependence between these functions $E_j(g_c)$ and the Cauchy kernel $C(g_c)$. In fact, we will want to reverse this process and define the $E_j(g_c)$ first, then define C via (4.1). This will require that we find an independent method for describing the $E_j(g_c)$. In fact, notice that form (4.3) and the fact that C is invariant under Ad K imply: For $g_0 \in G$,

$$\begin{aligned} E_j(g_0 g_c) &= \int_G C(g^{-1}(g_0 g_c)) D_j(g) dg \\ &= \int_G C((g_0^{-1}g)^{-1} g_c) D_j(g) dg \\ &= \int_G C(g^{-1} g_c) D_j(g_0 g) dg \\ &= D_j(g_0) E_j(g_c). \qquad (4.4) \end{aligned}$$

For k ε K,

$$E_j(g_c k) = \int_G C(g^{-1}g_c k) D_j(g) dg$$

$$= \int_G C(kg^{-1}g_c) D_j(g) dg$$

$$= \int_G C((gk^{-1})^{-1}g_c) D_j(g) dg$$

$$= \int_G C(g^{-1}g_c) D_j(gk) dg$$

$$= E_j(g_c) D_j(k). \qquad (4.5)$$

To draw the consequences of (4.4)-(4.5), introduce an orthonormal basis (Ψ_m), $1 \leq j \leq c(j)$, for H_j. Set:

$$d_j^{mm'}(g) = \langle \Psi_m | D_j(g)(\Psi_{m'}) \rangle$$

$$e_j^{mm'}(g) = \langle \Psi_m | E_j(g)(\Psi_{m'}) \rangle.$$

Then, (4.4) implies

$$e_j^{mm'}(gg_c) = \sum_{m''} d_j^{mm''}(g) e_j^{m''m'}(g_c), \qquad (4.6)$$

while (4.5) implies

CAUCHY INTEGRALS ON LIE GROUPS   101

$$e_j^{mm'}(g_c k) = \sum_{m''} e_j^{mm''}(g_c) d_j^{m''m}(k) \qquad (4.7)$$

This process can be reversed: Given sets of functions $e_j^{mm'}(g_c)$ satisfying (4.6)-(4.7), operators $E_j(g_c)$ can be defined satisfying (4.4)-(4.5), hence also a Cauchy kernel $C(g_c)$ defined using (4.1). In fact, notice that the general procedure sketched in Section 1 gives functions on G having the required transformation properties. In the next section we will show how the general procedure applies to the case $G_c = SO(3,C)$.

5. GROUP-THEORETIC MEANING OF THE INTEGRAL REPRESENTATIONS FOR THE LEGENDRE FUNCTIONS OF THE SECOND KIND

As we have seen in Section 4, the properties of the Cauchy kernel for a real Lie group are closely related to the properties to the "second kind" functions associated with the matrix elements of the representations of the groups. In particular, if we had an independent method for defining these "second-kind" functions, we would also have, at least in principle, a way to define the Cauchy kernels. The clue to this independent definition is the well-known integral representation for Legendre functions:

$$P_j^m(z) = \frac{\Gamma(j + m + 1)}{2\pi\Gamma(j)} \int_0^{2\pi} [z + (z^2-1)^{1/2} \cos\theta]^j \cos m\theta \, d\theta \qquad (5.1)$$

$$Q_j^m(z) = e^{im\pi} \frac{\Gamma(j + 1)}{\Gamma(j - m + 1)} \int_0^\infty [z + (z^2-1)^{1/2} \cos\theta]^{-j-1} \cosh(m\theta) \, d\theta \qquad (5.2)$$

Notice that it appears that one can be obtained from the other, at least formally and modulo normalizing factors by substituting $\theta \to -i\theta$. In [4] we have shown how the integral expansion for the Legendre functions themselves arise from the action of SL(2,R) on the unit circle in the complex plane. We will now show how this construction can be modified to give (5.2). This modification can also be interpreted by general principles, and this interpretation will provide us with a method to generalize the construction to more

CAUCHY INTEGRALS ON LIE GROUPS

complicated groups.

Let H be the a vector space of complex-valued functions of a real variable θ. Let $\underset{\sim}{G}$ be the Lie algebra (of SL(2,R)) generated by elements X, Y, Z, with:

$$[Z, X] = Y, [Z, Y] = X$$

$$[X, Y] = -Z. \qquad (5.3)$$

Consider the following one-parameter representation $\lambda \to D_\lambda$ of representations of $\underset{\sim}{G}$ by operators on H:

$$D_\lambda(Z) = \frac{d}{d\theta}; \quad D_\lambda(X) = \sinh\theta \frac{d}{d\theta} + \lambda \cosh\theta$$

$$D_\lambda(Y) = \cosh\theta \frac{d}{d\theta} + \lambda \sinh\theta$$

The eigenvectors of $D_\lambda(Z)$ are the functions:

$$\Psi_m = e^{m\theta}$$

We will show that the "matrix elements"

$$e(m, m', t, \lambda) = \int_{\infty}^{\infty} e^{m'\theta} \exp(tD_\lambda(X))(e^{m\theta}) d\theta$$

$$(5.4)$$

bear a close relation to the expressions (5.2).

The first step in the calculation of (5.4) is to use the trick mentioned in [4] and write:

$$D_\lambda(X) = (\sinh \theta)^{-\lambda} (\sinh \theta \tfrac{d}{d\theta})(\sinh \theta)^\lambda, \tag{5.5}$$

hence:

$$\exp(tD_\lambda(X)) = (\sinh \theta)^{-\lambda} \exp(t \sinh \theta \tfrac{d}{d\theta}) (\sinh \theta)^\lambda \tag{5.6}$$

To compute (5.6), let us solve for the orbits of the one-parameter group generated by the vector-field $\sinh \theta \tfrac{d}{d\theta}$:

$$\frac{d\theta}{dt} = \sinh \theta \quad \text{or} \quad \frac{d\theta}{\sinh \theta} = dt,$$

or

$$\frac{e^\theta - 1}{e^\theta + 1} = e^{t+c}.$$

Suppose that, for $t = 0$, $\theta = \theta_0$. Then,

$$e^c = \frac{e^{\theta_0} - 1}{e^{\theta_0} + 1}.$$

CAUCHY INTEGRALS ON LIE GROUPS   105

Hence

$$e^\theta = \frac{1 + e^{t+c}}{1 - e^{t+c}} = \frac{e^{\theta_0} + 1 + e^t(e^{\theta_0} - 1)}{e^{\theta_0} + 1 - e^t(e^{\theta_0} - 1)} \ .$$

In particular,

$$\exp(t \sinh \theta \, \tfrac{d}{dt})(e^\theta) = \frac{e^\theta + 1 + e^t(e^\theta - 1)}{e^\theta + 1 - e^t(e^\theta - 1)}$$

$$(5.7)$$

$$\exp(t \sinh \theta) \, \tfrac{d}{d\theta} (\sinh \theta)$$

$$= \frac{2(e^\theta + 1)(e^t(e^\theta - 1)}{(e^\theta + 1)^2 - (e^t(e^\theta - 1)^2)}$$

$$= \frac{2e^t(e^{2\theta} - 1)}{e^{2\theta}(1 - e^{2t}) + 2e^\theta(1 + e^{2t}) + (1 - e^{2t})}$$

$$= \frac{2 \sinh \theta}{e^\theta(\sinh t) + 2 \cosh t + \sinh t \, e^{-\theta}}$$

$$= \frac{\sinh \theta}{\sinh t \cosh \theta + \cosh t} \qquad (5.8)$$

This gives a main result:

$$\exp(tD_\lambda(X))(e^{m\theta}) = \left[\frac{e^\theta + 1 + e^t(e^\theta-1)}{e^\theta + 1 - e^t(e^\theta-1)}\right]^m$$

$$(\sinh t \cosh \theta + \cosh t)^{-\lambda} \qquad (5.9)$$

Note that this explains the origin of both (5.1) and (5.2) (up to constant factors) as matrix elements of representations of the group generated by the Lie algebra $\underset{\sim}{G}$.

$$\int_{-\infty}^{\infty} e^m \exp(tD_\lambda(X))(1) d\theta$$

$$= 2\int_0^\infty (\sinh t \cosh \theta + \cosh t)^{-\lambda} \cosh(m\theta) d\theta. \qquad (5.10)$$

To obtain (5.1), we must analytically continue θ from real to pure imaginary values. This can be done most understandably in the following way: Set

$$u = e^\theta.$$

Then,

$$D_\lambda(Z) = \frac{du}{d\theta}\frac{d}{du} = u\frac{d}{du}$$

$$D_\lambda(X) = \frac{1}{2}(u^2 - 1)\frac{d}{du} + \frac{\lambda}{2}(u + u^{-1})$$

CAUCHY INTEGRALS ON LIE GROUPS 107

$$D_\lambda(Y) = \frac{1}{2}(u^2 + 1)\frac{d}{du} + \lambda(u - u^{-1}) \qquad (5.11)$$

This is clearly the Lie algebra of SL(2,R), considered as a transformation group on the complex u-plane via linear fractional transformations. Consider (5.11) as defining the representation D_λ of $\underset{\sim}{G}$ by differential operators on the space of functions that are holomorphic in the complex u-plane, minus the origin, with at most a pole at the origin. We can then define an inner product for these functions by integrating over the positive real axis (which corresponds to (6.9)) or over the unit circle in the complex u-plane. The measure chosen for the positive real axis du/u is that which is invariant under the group generated by the vectorfield u d/du, of which it is an orbit. The measure to be used for the unit circle is that invariant under u d/du, which is not an element of $\underset{\sim}{G}$, but an element of $\underset{\sim}{G}_c$. In the next section, we shall present a generalization of this idea to other groups. Returning to the derivation of (5.1), notice that, with the inner product determined by integrating functions in the u-plane over the unit circle $u = e^{i\theta}$, $0 \leq \theta \leq 2\pi$, the matrix elements become:

$$\langle e^{im\theta} | \exp(tD_\lambda)(1) \rangle$$

$$= \int_0^{2\pi} e^{-im\theta}(\sinh t \cos \theta + \cosh t)^{-\lambda} d\theta$$

$$= \int_0^{2\pi} (\sinh t \cos \theta + \cosh t)^{-\lambda} \cos m\theta \, d\theta, \tag{5.12}$$

which gives (6.1).

Let us return to the study of the representation D_λ of $\underset{\sim}{G}$. For $g \in \exp(\underset{\sim}{G})$, i.e. $g = \exp(W)$, with $W \in \underset{\sim}{G}$, set $D_\lambda(g) = \exp(D_\lambda(W))$. For m, m' integers, set:

$$e(m', m, g) = \langle e^{m'\theta} | D_\lambda(g)(e^{m\theta}) \rangle$$

$$\equiv \int_{-\infty}^{\infty} e^{m'\theta} D_\lambda(g)(e^{m\theta}) d\theta.$$

As explained in Section 4, we are interested in finding out how these functions on g transform under left translation by elements of G. Now, for $g' \in G$,

$$e(m', m, g_0^{-1}g) = \langle e^{m'\theta} | D_\lambda(g_0^{-1}) D_\lambda(g) e^{m\theta} \rangle$$

CAUCHY INTEGRALS ON LIE GROUPS

$$= \langle D_\lambda(g_0^{-1})^* e^{m'\theta} | D_\lambda(g) e^{m\theta} \rangle,$$

where * denotes the adjoint operator with respect to the form $\langle \ | \ \rangle$. Now,

$$D_\lambda(X)^* = -\frac{d}{d\theta} \sinh \theta + \lambda^* \cosh \theta$$

$$= -\sinh \theta \frac{d}{d\theta} + (\lambda^* - 1) \cosh \theta$$

$$= -D_{1-\lambda^*}(X)$$

$$D_\lambda(Y) = -\frac{d}{d\theta} \cosh \theta + \lambda^* \sinh \theta$$

$$= -D_{1-\lambda^*}(Y)$$

Then,

$$D_\lambda(W)^* = -D_{1-\lambda^*}(W) \quad \text{for all} \quad W \in \underset{\sim}{G}, \text{ and}$$

$$\text{if} \quad g_0 = \exp(W),$$

$$D_\lambda(g_0^{-1})^* = D_{1-\lambda^*}(g_0),$$

hence:

$$e(m', m, g_0^{-1}g) = \langle D_{1-\lambda^*}(g_0)(e^{m'\theta}) | D_\lambda(g) e^{m\theta} \rangle$$

The key question is now: For what values of λ,

and m' is $D_{1-\lambda}{}^*(g_0)(e^{m'\theta})$ a linear combination of a *finite* number of the $e^{m\theta}$'s. This will happen when λ is such that $D_{1-\lambda}{}^*$ acting on H leaves invariant a finite dimensional subspace consisting of a finite number of the $e^{m\theta}$'s. In fact, notice that

$$D_{1-\lambda}{}^*(X + Y) = e^\theta \frac{d}{d\theta} + (1 - \lambda^*)e^\theta,$$

hence

$$D_{1-\lambda}{}^*(X + Y)(e^{\ell\theta}) = [\ell + (1 - \lambda^*)]e^{(\ell+1)\theta}$$

Then, if ℓ is real, this will be zero if $\lambda = \ell+1$. If this condition is satisfied, we also have:

$$D_{1-\lambda}{}^*(X - Y)(e^{-\ell\theta}) = 0.$$

If ℓ is an integer, the functions $e^{-\ell\theta}, \ldots e^{\ell\theta}$ are then transformed among themselves by $D_{-\ell}(\underset{\sim}{G})$, hence $D_{-\ell}(G)$, acting on this subspace, is globally defined over all G. Suppose

$$D_{-\ell}(g)(e^{m'\theta}) = \sum_{-\ell \leq m \leq \ell} d(\ell, m, m', g)e^{m\theta},$$

i.e. the d-functions are the matrix element functions of the finite dimensional representations of SL(2,R) (hence also ℓ when analytically continued, the matrix element functions of the finite dimensional representations of SO(3,R)). Hence, for $\lambda = \ell+1$,

$$e(m', m, g_0^{-1}g) = <D_{-\ell}(g_0)e^{m'\theta}|D_{\ell+1}(g)e^{m\theta}>$$

$$= \sum_{-\ell \leq m'' \leq \ell} d(\ell, m'', m'g_0)<e^{m''\theta}|D_{\ell+1}(g)e^{m\theta}>$$

$$= \sum_{m''} d(\ell, m'', m', g_0)e(m'', m, g).$$

Thus, the functions $g \to e(m', m, g)$ on G, for these values of λ (and for m', m suitably restricted) transform under left translation via a finite dimensional representation of G. However, notice that they do not similarly transform via right translation, although they do transform like a finite dimensional representation when right translated by the subgroup exp(tZ), which we take to be the subgroup K. Notice also that $\lambda = \ell+1$ is the value of λ with the property that $e(0, \theta, \exp(tX)) = Q_\ell(\cosh t)$, where Q_ℓ is the classical "Legendre function of the second kind." In summary, we have succeeded in tying together the general procedure sketched in the introduction for defining "matrix element functions of the second kind" and the classical definition of the Legendre functions of the second kind via integral representations. We now turn to a general description of the method used here.

6. MATRIX ELEMENT FUNCTIONS OF THE SECOND KIND FOR GENERAL GROUPS

Let G be a real noncompact, semisimple Lie group, with L a subgroup, such that G/L is a "boundary homogeneous space" of G, [2]. Let $G \subset G_c$, be the complexification of G, L_c the complexification of G, L_c the complexification of L. Suppose that also $L_c \subset G_c$. Then, G/L is naturally contained in G_c/L_c, a complex manifold that may be thought of as the "complexification" of G/L.

Suppose a one parameter family of representation $D_\lambda(G)$ of representations of G by operators in a vector space H of functions on G/L is defined as follows:

CAUCHY INTEGRALS ON LIE GROUPS

$$D_\lambda(g)(\Psi)(p) = m(g, p)^\lambda \Psi(g^{-1}p), \qquad (6.1)$$

where p denotes a point of G/L, $g \in G$, $\Psi \in H$, and where $m(g, p)$ is a multiplier system for the action of G on G/L [4].

Now, let us suppose that everything can be "analytically continued." H is supposed to consist of functions that are holomorphic on G_c/L_c, with possible singularities. Assume $m(g, p)$ can be continued holomorphically to $g \in G_c$, $p \in G_c/L_c$, with the possibility of singularities again. Assume finally that H is mapped into itself under translation by G_c, so that (6.1) makes sense for $g \in G_c$, $p \in G_c/L_c$.

Suppose that N is another real submanifold of G_c/L_c. (It may be G/L itself, or possibly the orbit of another real form of G_c.) One can define an inner product in H by the rule:

$$\langle \Psi | \Psi' \rangle = \int_N \Psi(p)^* \Psi'(p) dp$$

Here, dp is a volume element on H. Of course, this integral might not be finite, due either to the singularities of the elements of H, or to the possible non-compactness of N: This is the sort of inner product mentioned in the introduction.

Suppose N and dp can be chosen so that:

$$D_\lambda(g^{-1})^* = D_{\lambda'}(g) \quad \text{for} \quad g \in G$$

where * denotes the adjoint operator relative to the form $< \ | \ >$, and $\lambda \to \lambda'$ is a mapping of the parameter space into itself. Then, for certain values of λ, $D_{\lambda'}(G)$ can be expected to leave invariant finite dimensional subspaces of H. These subspaces can be used, as indicated in the introduction, to construct "matrix element functions of the second kind."

BIBLIOGRAPHY

1. M. Andrews and J. Gunson, Properties of Local Representations of the Rotation Group, J. Math. Phys. 5, 1391 (1964).
2. R. Hermann, Lie Groups for Physicists, W. A. Benjamin, 1966.
3. _____, Some Properties of Representations of Non-compact Groups, in "High Energy Physics and Elementary Particles," ICTP, Trieste, 1965.
4. _____, Analytic Continuation of Group Representations, Comm. Math. Phys., Part I, 2, 251-270 (1966); Part II, 3, 53-74 (1966); Part III, 3, 75-97 (1966); Part IV, 5, 131-156 (1967); Part V, 5, 157-190 (1967).
5. L. K. Hua, Harmonic Analyses of Functions of Several Complex Variables in the Complex Domains, Translations of Mathematical monographs, Vol. 6, American Math. Soc., Providence, 1963.
6. E. M. Stein and S. Wainger, Analytic Properties of Expansions, and some variants of the Parseval-Plancherel formulas, Arkiv for Math., 5 (1965), 553-567.

CHAPTER IV

DEFORMATION OF THE FOURIER INTEGRAL ON GROUPS FROM COMPACT TO NON-COMPACT GROUPS

1. INTRODUCTION

The main mathematical problem discussed in this series of papers may be defined as that of describing the relation between the Fourier expansion of functions on different subgroups of a given Lie group G. Our main aim is to develop methods of "analytic continuation" or "deformation" of the subgroups.

Two such methods may be discerned. The first (introduced in Chapter I, Section 8), using a variant of the Inonu-Segal-Wigner "contraction" idea, [1], seeks to describe how the Fourier expansion approaches a limit as sequences of subgroups approach limits. The second, described in Chapters 2 and 3, seeks to relate Fourier expansions on different real-forms of a given complex group by means of Cauchy integral formulae.

In this book, we present further work on the first method. We concentrate on the simplest examples, $G = SO(2,1)$, or $SO(3,1)$--the problem seems to involve new, intricate, and delicate problems in the theory of analysis and generalized functions on Lie groups, and before working on the general theory it is necessary to explicitly work out representative examples. Since the material given in Chapter I may be regarded as an introduction to the problem, we shall proceed immediately to the details.

2. LIMITING RELATIONS FOR SUBGROUPS OF $SO(2,1)$

Let G be the group $SO(2,1)$ (the Lorentz group in two space variables). Let $\underset{\sim}{G}$ denote its Lie algebra. It is generated by elements X_1, X_2, X_3 satisfying:

DEFORMATION OF THE FOURIER INTEGRAL

$$[X_1, X_2] = X_3; \quad [X_1, X_3] = -X_2$$

$$[X_2, X_3] = -X_1. \qquad (2.1)$$

Let K be the subgroup $\theta \to \exp(\theta X_1)$. It is isomorphic to SO(2,R). Let $t \to g(t)$ be the one-parameter, non-compact subgroup given by:

$$g(t) = \exp(tX_3).$$

For each t, let:

$$K_t = \exp(g(t))(K) = g(t)Kg(-t)$$

Let L be the one-parameter subgroup

$$a \to \exp(a(X_1 + X_3))$$

Then, the limit as $t \to \infty$ of the subgroups K_t is equal to the subgroup L. (This sort of limit is described in [1], Chapter XI.)

Suppose that $g \to f(g)$ is a complex-valued, C^∞ function on G of compact support. Then, restricting f to each subgroup K_t, we can expand it in a Fourier series:

$$f(k_t) = \sum_{j=-\infty}^{\infty} a_j t e^{ij\theta(k_t)} \qquad (2.2)$$

for $k_t \in K$.

with

$$a_j^t = \int_{K_t} f(k_t) e^{-ij\theta(k_t)} dk_t \qquad (2.3)$$

Here, $\theta(k_t)$ is the angle of rotation of k_t, i.e.

$$k_t = g(t) \exp(\theta(k_t) X_1) g(-t) \qquad (2.4)$$

Using the commutation relations (2.1), one shows that:

$$k_t = \exp(\theta(k_t)(\cosh t X_1 + \sinh k t X_3)) \qquad (2.5)$$

Then, by definition,

$$\int_{K_t} f(k_t) dk_t = \frac{1}{2\pi} \int_{\pi}^{\pi} f(g(t) \exp(\theta X_1) g(-t)) d\theta. \qquad (2.6)$$

Suppose k_t is a sequence of elements of G, with each $k_t \in K_t$, such that:

$$\lim_{t \to \infty} k_t = \ell \in L. \qquad (2.7)$$

For example, if

$$\ell = \exp(\tfrac{a}{2}(X_1 + X_3)),$$

from (2.5) we see that (2.7) will be satisfied if:

$$\theta(k_t) = e^{-t}a. \qquad (2.8)$$

With (2.7) satisfied, hence the left hand side of (2.2) converging as $t \to \infty$ to $f(\ell)$, suppose we investigate the right hand side of (2.2). One would expect that it converges, as $t \to \infty$, to the Fourier decomposition of the function $\ell \to f(\ell)$ on L. Since L is an abelian group, this is essentially just the ordinary Fourier integral decomposition of the function

$$a \to f(\exp(a \tfrac{(X_1 + X_3)}{2})).$$

One might hope to apply the method introduced in Chapter I (for the more complicated case $G = SO(3,1)$, $K = SO(3,R)$). Recall that this involved introducing an irreducible, unitary representation of G on a Hilbert space H that decomposes under K into irreducible representations, with each representation occurring precisely once in this decomposition. The method then proceeds by rewriting the right hand side of (2.2) as a trace of an operator on H. However, as we saw in Chapter I, describing the limit as $t \to \infty$ of these traces involved subtle questions of convergence of operators in H, that, in fact, were not completely resolved in that section. In this section, we will present a simpler, more classical method, for treating the limit as $t \to \infty$, using the partial sums of (2.2) and the classical Dirichlet Fourier-series kernel. Define:

$$f_n(k_t) = \sum_{j=-n}^{n} a_n t e^{in\theta(k_t)} \qquad (2.9)$$

Then, using (2.3), we have:

$$f_n(k_t) = \int_{K_t} \frac{\sin((n+\tfrac{1}{2})(\theta(k_t)-\theta(k_t')))}{\sin(\tfrac{1}{2}(\theta(k_t)-\theta(k_t')))} f(k_t') dk_t'$$

$$= e^t \int_{K_t} \frac{\sin((n+\tfrac{1}{2})(\theta(k_t)-\theta(k_t')))}{e^t(\sin(\tfrac{1}{2}(\theta(k_t)-\theta(k_t'))))} f(k_t') dk_t'$$

Suppose that (2.8) holds. Then, using (2.5),

DEFORMATION OF THE FOURIER INTEGRAL 123

$$f_n(k_t) = \frac{e^t}{2\pi} \int_{-\pi}^{\pi} \frac{\sin((n+\tfrac{1}{2})(e^{-t}a-\theta))}{e^t \sin(\frac{e^{-t}a-\theta}{2})}$$

$$f(\exp(\theta(\cosh t X_1 + \sinh t X_3)))d\theta$$

Change variables of integration: $\theta \to e^{-t}\theta'$. Then,

$$f_n(k_t) = \frac{1}{2\pi} \int_{-\pi e^t}^{\pi e^t} \frac{\sin((n+\tfrac{1}{2})e^{-t}(a-\theta))}{e^t \sin(\frac{e^{-t}}{2}(a-\theta))}$$

$$f(\exp(\theta e^{-t}(\cosh t X_1 + \sinh t X_3)))d\theta \quad (2.10)$$

We know that:

$$\lim_{t\to\infty} \lim_{n\to\infty} f_n(k_t) = f(\ell), \quad (2.11)$$

with

$$\ell = \exp(\tfrac{a}{2}(X_1 + X_3)).$$

(2.10) suggests that we rewrite the limit on the left-hand side of (2.11) substituting

$$N = (n + \tfrac{1}{2})e^{-t} \quad (2.12)$$

Then,

$$f(\ell) = \lim_{t \to \infty} \lim_{n \to \infty} \frac{1}{2\pi} \int_{-\pi e^t}^{\pi e^t} \frac{\sin(N(a-\theta))}{e^t \sin(\frac{e^{-t}}{2}(a-\theta))}$$

$$\times f(\exp(\theta e^{-t}(\cosh t X_1 + \sinh t X_3)))d\theta$$

(2.13)

Now, use the Fourier integral theorem to express $f(\ell)$:

$$f(\ell) = \lim_{N \to \infty} \frac{1}{2\pi} \int_{-N}^{N} \left(\int_{-\infty}^{\infty} f(\exp \tfrac{b}{2}(X_1 + X_3)) \right.$$

$$\left. e^{-iba'} db \right) e^{iaa'} da'$$

$$= \lim_{N \to \infty} \frac{1}{2\pi} \int_{-\infty}^{\infty} \frac{2 \sin N(\theta - a)}{(\theta - a)} f(\exp(\tfrac{\theta}{2}(X_1 + X_3)))d\theta.$$

(2.14)

We see that we can also obtain (2.14) formally from (2.13) by interchanging limits in (2.13), and then letting $t \to \infty$. This we have made precise the formal process whereby the right hand side of (2.2) goes into the Fourier representation of $\ell \to f(\ell)$ as $t \to \infty$. Presumably, one can work directly to show that these limits can be interchanged (i.e. without using the Fourier integral

DEFORMATION OF THE FOURIER INTEGRAL 125

theorem to express (2.14)), and thus *prove* the Fourier integral expansion theorem from the Fourier series expansion theorem. This is a refinement of the argument that we will not go into here.

In summary, we have described a method for passing from the Fourier expansion of a function on K_t to the Fourier expansion of the function on the "limit" group L. This method applies to much more general situations: as an example, we now turn to the case of physical interest (in connection with the behavior of the partial wave analysis of the scattering amplitude in the region where the "little group" changes from compact to noncompact), namely $G = SO(3,1)$.

3. DIRECT PASSAGE FROM A SUM TO AN INTEGRAL

The method used in Section 2 is difficult to generalize to more general situations, mainly because it is difficult to work with the sum of the characters of irreducible, finite dimensional representations of greater complexity than those of $SO(2,R)$. We will now present another method for treating the case of Section 2, which is more readily generalizable. As a defect, it is probably harder to make rigorous, depending as it does on a difficult-to-estimate replacement of an integral by an approximating Riemann sum. Suppose

$$G, K, t \to g(t), X_1, L, Y_3, Y_2, g \to f(g)$$

are as in Section 2. Write:

$$\ell = \lim_{t \to \infty} g(t)\exp(\theta e^{-t} X_1) g(-t) \equiv \lim_{t \to \infty} k_t$$

$$\equiv \lim_{t \to \infty} g(t) k(t) g(-t).$$

Then,

$$f(\ell) = \lim_{t \to \infty} f(k_t)$$

$$= \lim_{t \to \infty} \lim_{n \to \infty} \frac{1}{2\pi} \int_{-\pi}^{\pi} \left(\sum_{j=-n}^{n} e^{ij(e^{-t}\theta - \theta')} \right) f(g(t)\exp(\theta' X_1) g(-t)) d\theta'$$

$$= \lim_{t \to \infty} \lim_{N \to \infty} \frac{1}{2\pi} \int_{-\pi N}^{\pi N} \left(\sum_{j=-e^{t}N}^{e^{t}N} e^{ij e^{-t}(\theta - \theta')} e^{-t} \right) f(g(t)\exp(\theta' e^{-t} X_1) g(-t)) d\theta' \qquad (3.1)$$

DEFORMATION OF THE FOURIER INTEGRAL

Now,

$$\sum_{j=-e^{t}N}^{e^{t}N} e^{ije^{-t}(\theta-\theta')} e^{-t}$$

$$= \sum_{h=-N,-N+e^{-t},-N+2e^{-t},\ldots,N} e^{ih(\theta-\theta')} e^{-t}$$

Notice that this is an approximating Riemann sum for the integral:

$$\int_{-N}^{N} e^{ih(\theta-\theta')} dh. \qquad (3.2)$$

Thus, if the limits can be interchanged in (3.1), and if the replacement of the sum by (3.2) in this limit is legitimate, then

$$f(\ell) = \frac{1}{2\pi} \lim_{N\to\infty} \int_{-N}^{N} e^{ij\theta} \left(\int_{-\infty}^{\infty} f(\exp(\frac{\theta'}{2}(X_1 + Y_2))) e^{-i\theta'} d\theta' \right) d\theta,$$

which is just the Fourier integral expansion.

4. DEFORMATION OF THE FOURIER EXPANSION FROM SO(3,R) TO E(2)

Consider now the case: $G = SO(3,1)$, $K = SO(3,R)$. $\underset{\sim}{G}$ now has a basis (X_i, Y_i), $1 \leq i, j, k \leq 3$, satisfying:

$$[X_i, Y_j] = \varepsilon_{ijk} Y_k$$
$$[X_i, X_j] = \varepsilon_{ijk} X_k$$
$$[Y_i, Y_j] = -\varepsilon_{ijk} X_k \qquad (4.1)$$
$$1 \leq i, j, k \leq 3,$$

Set

$$g(t) = \exp(tY_3) \qquad (4.2)$$

$\underset{\sim}{K}$ is spanned by X_1, X_2, X_3 \qquad (4.3)

For each $t \geq 0$, let K_t be the following subgroup of G:

$$K_t = g(t) K g(-t) \qquad (4.4)$$

Suppose that L is a subgroup of G such that

$$\lim_{t \to \infty} K_t = L, \qquad (4.5)$$

DEFORMATION OF THE FOURIER INTEGRAL

as defined in [1, Chapter 11]. In fact, one can show that L is the subgroup whose Lie algebra is given as follows:

$$\underline{L} = \{X_3, X_1 + Y_2, X_2 - Y_1\} \qquad (4.6)$$

Recall that $\underline{L}$ is isomorphic to the group of rigid motions in the Euclidean plane, which we denote by E(2).

Let us adopt the Euler angle coordinates $(\theta_1, \theta_2, \theta_3)$ for elements k of K.

$$k = \exp(\theta_1 X_3)\exp(\theta_2 X_1)\exp(\theta_3 X_3) \qquad (4.7)$$

Then, as is well-known,

$$\int_K f(k)dk = \frac{1}{8\pi^2} \int_0^{2\pi} \int_0^{\pi} \int_0^{2\pi} f(\exp(\theta_1 X_3)$$
$$\exp(\theta_2 X_1)\exp(\theta_3 X_3))\sin\theta_2 d\theta_1 d\theta_2 d\theta_3$$
$$(4.8)$$

For $k \in K$, and each positive real number t, define

$$k(t) = \exp(\theta_1 X_3)\exp(e^{-t}\theta_2 X_1)\exp(\theta_3 X_3),$$

where $k \in K$ is given by (4.7). Notice that:

$$\lim_{t \to \infty} g(t)k(t)g(-t) = \exp(\theta_1 X_3)$$

$$\exp(\frac{\theta_2}{2}(X_1 + Y_2))\exp(\theta_3 X_3). \qquad (4.9)$$

Notice that the right hand side of (4.9) is an element of L. Let us call it ℓ. Now,

$$\int_K f(g(t)kg(-t))dk$$

$$= \frac{1}{8\pi^2} \int_0^{2\pi} \int_0^{\pi} \int_0^{2\pi} f(g(t)\exp(\theta_1 X_3)$$

$$\exp(\theta_2 X_1)\exp(\theta_3 X_3))\sin \theta_2 d\theta_1 d\theta_2 d\theta_3$$

$$= \frac{1}{8\pi^2} \int_0^{2\pi} \int_0^{\pi} \int_0^{2\pi} f(\exp(\theta_1 X_3)\exp(\theta_2(\cosh t X_1$$

$$+ \sinh t Y_2))\exp(\theta_3 X_3)\sin \theta_2 d\theta_1 d\theta_2 d\theta_3$$

$$= \frac{e^{-2t}}{8\pi^2} \int_0^{2\pi} \int_0^{\pi e^t} \int_0^{2\pi} f(\exp(\theta_1 X_3)$$

$$\exp(\theta_2 e^{-t}(\cosh t X_1 + \sinh t Y_2))\exp(\theta_3 X_3)$$

$$\times e^t \sin(e^{-t}\theta_2)d\theta_1 d\theta_2 d\theta_2$$

This calculation shows that:

$$\lim_{t \to \infty} e^{2t} \int_K f(g(t)kg(-t))dk = \int_L f(\ell)d\ell \qquad (4.10)$$

DEFORMATION OF THE FOURIER INTEGRAL 131

where $d\ell$ is the biinvariant volume element on L.

Let us copy the procedure used in Section 2. Suppose that $g \to f(g)$ is a C^∞ function on G with compact support. Then,

$$f(g(t)k(t)g(-t))$$
$$= \sum_{j=0}^{\infty} \text{trace}(\hat{f}_j{}^t D_j(k(t)))(2j+1),$$

with

$$\hat{f}_j{}^t = \int_K f(g(t)k'g(-t))D_j(k'^{-1})dk', \qquad (4.11)$$

where $k \to D_j(k)$ is the spin j, irreducible unitary representation of $K = SO(3,R)$. Now, set

$$f_n(k, t) = \sum_{j=0}^{n} \text{trace}(\hat{f}_n D_j(k(t)))2j + 1$$
$$= \int_K \alpha_n(k(t), k')f(g(t)k'g(-t))dk',$$

with α_n defined as:

$$\alpha_n(k(t), k') = \sum_{j=0}^{n} \text{trace}(D_j(k(t)k'^{-1})(2j+1)$$
$$(4.12)$$

Let us rewrite (4.12) in a form that will facilitate replacing it by a definite integral. First, set

$$n = Ne^t. \qquad (4.13)$$

Then,

$$\alpha_n(k(t), k'(t)^{-1})$$

$$= \sum_{j=0}^{Ne^t} \text{trace}(D_j(k(t)k'(t)^{-1})(2j+1)$$

= after substituting $h = je^{-t}$,

$$\sum_{h=0, e^{-t}, 2e^{-t}\ldots N} \text{trace} D_{he^t}(k(he^t)$$

$$k'(he^t)^{-1})(2he^t + 1)$$

$$= e^{2t} \sum_{h=0, e^{-t}, \ldots, N} \text{trace} D_{he^t}(k(he^t)$$

$$k'(he^t)^{-1})(e^{-t})(2h + e^{-t}) \qquad (4.14)$$

As is well-known, $D_j(k)$ can be realized as $(2j + 1) \times (2j + 1)$ unitary matrices, that we denote by

$$D_j^{mm'}(k), \quad -j \leq m, m' \leq j. \qquad (4.15)$$

These matrices can be defined as follows: There is a Hilbert space H, a unitary, irreducible

DEFORMATION OF THE FOURIER INTEGRAL   133

representation $g \to D(g)$ of G by operators on H.
H has a discrete orthonormal basis, labelled,
using the Dirac notations, by

$$|m, j>, \quad 0 \leq j \leq \infty, \quad -j \leq m \leq j \qquad (4.16)$$

For fixed j, the states $|-j, j>$, $|-j+1, j>$,...,
$|j, j>$ are transformed among themselves by D(K)
via the spin j-representation, and such that:

$$D(\exp(\theta X_3))|m, j> = e^{im\theta}|m, j> \qquad (4.17)$$

Then, we can define the matrix elements indicated
in (3.17) by:

$$D_j^{mm'}(k) = <m, j|D(k)|m', j> \qquad (4.18)$$

Thus,

$$\alpha_n = e^{2t} \sum_{h=0, e^{-2},\ldots,N} \sum_m D_{he^t}^{mm}(k(he^t))$$

$$k'(he^t)^{-1}(e^{-t})(2h + e^{-t}) \qquad (4.19)$$

Now, of course, for (4.19) to make sense, he^t must
be an integer. However, we know that the functions
$D_{he^t}^{mm}(k)$ can in fact be defined for all real h,

with integer m's. With this understanding, we can regard (4.19) as an approximation for an integral, as follows:

$$e^{-2t}\alpha_n \sim \int_0^N \sum_m D^{mm}_{he^t}(k(ht)k'(ht)^{-1})(2h+e^{-t})dh \quad (4.20)$$

With:

$$\ell = \lim_{t \to \infty} g(t)k(t)g(-t) \, \varepsilon \, L,$$

and

$$\ell' = \lim_{t \to \infty} g(t)k'(t)g(-t), \quad (4.21)$$

define:

$$D_h^{mm'}(\ell) = \lim_{t \to \infty} D^{mm'}_{he^t}(k(he^t)) \quad (4.22)$$

Then, we see that (at least formally) as $n, t \to \infty$,

$$f(\ell) = \lim_{N \to \infty} \int_0^N \int_L \sum_m D_h^{mm}(\ell\ell'^{-1})$$

$$f(\ell')2h d\ell' dh \quad (4.23)$$

Since (4.23) may be expected to hold for a class

DEFORMATION OF THE FOURIER INTEGRAL　　　　135

of functions on L that is dense in the class of continuous functions with compact support, it seems likely that (4.23) is *the* Fourier expansion of functions on L in terms of matrix elements of irreducible unitary representation. To see this, we should exhibit the left hand side of (4.22) as the matrix elements of such representations. We will present two methods for accomplishing this aim.

The first method utilizes the unitary representation $g \to D(g)$ of all of G. Now,

$$D^{mm'}_{he^t}(k(he^t)) = <m, he^t | D(k(he^t)) | m', he^t>$$

$$= <D(g(t)(|m, he^t>) | D(g(t)k(he^t)$$

$$g(-t)) | D(g(t) | m', he^t>$$

Now,

$$\lim_{t \to \infty} g(t)k(he^t)g(-t) = \ell_h, \qquad (4.24)$$

where ℓ_h is defined as follows:

$$\ell_h = \exp(\theta_1 X_3)\exp(\frac{\theta_2 h}{2}(X_1 + Y_2))\exp(\theta_3 X_3)$$

$$(4.25)$$

Suppose "states" of H can be defined by:

$$|m, h\rangle = \lim_{t \to \infty} D(g(t)|m, he^t\rangle \qquad (4.26)$$

Then, we have, at least in a formal way,

$$D_h^{mm'}(\ell) = \langle m, h|D(\ell_h)|m', h\rangle \qquad (4.27)$$

Of course, the limits (4.26) would not necessarily exist in the ordinary sense, but in "weak" sense (i.e. the inner products with a subspace of elements of H might be expected to converge). Thus, one might expect that the "states" $|m, h\rangle$ are not elements of H, but of the "Dirac space" built-up from H. (See [Chapter I] and [2, Part 6].)

Another approach is to work directly with the relations (4.22). They are, in fact, very reminiscent of the classical limit relations among Legendre and Bessel functions:

$$\lim_{j \to \infty} P_j^m(\cos \frac{\theta}{j}) = J_m(\cos \theta) \qquad (4.28)$$

Now, the P_j^m are related to matrix elements of representations $SO(3,R)$, the J_j are related to matrix elements of E(2). In [2, Part 5, Section 4], we have shown how (4.28) may be obtained from deformation-of-representation considerations. The method given there extends readily to show that (4.22) holds, with the left-hand side the matrix elements of representation of E(2).

BIBLIOGRAPHY

1. R. Hermann, Lie Groups for Physicists, W. A. Benjamin, New York, 1966.

2. _____, Analytic Continuation of Group Representations, Commun. Math. Phys. Part 1, 2, 251-270 (1966); Part 2, 3, 53-74 (1966); Part 3,3, 75-97 (1966); Part 5,5, 157-190 (1967); Part 6,6, 205-225 (1967).

CHAPTER V

PARTIAL WAVE ANALYSIS OF THE SCATTERING AMPLITUDE

In this chapter we will present a treatment of "partial wave analysis". In Chapter VI we will present a more general approach that is slightly more sophisticated from a group-theoretical viewpoint. However, the material in this chapter is closer to the traditional physicist's approach, and also throws some light on the problem of "kinematic singularities" in the study of the scattering amplitude.

1. SOME GENERAL PRINCIPLES OF GROUP REPRESENTATIONS THEORY

First, we will recall some general principles of the "physicist's approach toward group representation theory."

Let G be a group, and let $g \to D(g)$ be a representation of G by operators on a Hilbert space H. Let $\underset{\sim}{D}$ be the "Dirac space" associated with H. Let us suppose that, for each $g \in G$, the adjoint operator $D(g)^*: H \to H$ exists, i.e.

$$<\psi|D(g)\psi'> = <D(g)^*\psi|\psi>$$

for $\psi, \psi' \in H$.

Then, $D(g)$ can be extended to an operator: $\underset{\sim}{D} \to \underset{\sim}{D}$ by the following formula:

$$<D(g)\alpha|\psi> = <\alpha|D(g)^*\psi>$$

for $\psi \in H, \alpha \in \underset{\sim}{D}$

One readily verifies that this extension of $D(G)$ to operators on $\underset{\sim}{D}$ is also a representation of G.

Let $\underset{\sim}{G}$ be the Lie algebra of G. For $X \in \underset{\sim}{G}$, let us suppose that $\underset{\sim}{D}(X)$ can be defined by the following formula:

$$\underset{\sim}{D}(X)\psi = \frac{\partial}{\partial t} D(\exp(tX))\psi \Big|_{t=0}$$

and that $X \to \underset{\sim}{D}(X)$ is a Lie algebra representation of $\underset{\sim}{G}$ by operators on H. Further, suppose that D(G) can similarly be extended to give a representation of $\underset{\sim}{G}$ by operators on $\underset{\sim}{D}$.

These assumptions are satisfied when one constructs representations by means of the usual method [5,7] of induced representation -- vector bundle theory, choosing H as some space of suitably "smooth" cross-sections -- usually with restricted behavior at infinity of homogeneous vector bundles on coset spaces of G. In fact, it would be very useful to have available a much more complete theory describing realistically the relation between the "abstract", functional analysis view of group representations and the geometric view, but one cannot say that this exists at the moment. Hence, we will proceed in a somewhat loose way that will probably offend the sensibilities of those mathematicians who regard the nooks and crannies of contemporary functional analysis -- unrelated to interesting concrete examples -- as the most important objects to study.

A central idea is to "diagonalize a maximal abelian set of operators." The physicist thinks of this as "diagonalizing a maximal commuting set of observables" -- it is, in fact, one of the basic ideas that he carries over from his study of quantum mechanics. Of course, group representation theory and quantum mechanics have a very similar flavor, as has been long realized (although perhaps not sufficiently so from an educational point of view) since the classic works of Weyl and Wigner.

One way of implementing this can be described as follows: Let A be an abelian subgroup of G, and let C be the center of the universal enveloping algebra $U(G)$ of G, i.e. the elements of C are "Casimir operators" of G [5]. D can be extended to a representation of $U(G)$, and then

$$D(A), D(C)$$

form a "commuting set of operators on D." Let us postulate its "diagonalization" in the following form:

Suppose M is a space, with a measure, denoted by dp. Let $p \to \psi_p$ be a map: $M \to D$ such that:

$$D(a)(\psi_p) = \lambda(a, p)\psi_p$$
$$\underset{\sim}{D}(\Delta)(\psi_p) = \delta(\Delta, p)\psi_p \qquad (1.1)$$

for $a \in A$, $\Delta \in \underset{\sim}{C}$, $p \in M$.

(1.1) says that the "ψ_p are eigenvectors of the operators $D(A)$, $D(\underset{\sim}{C})$." Then λ and δ can be regarded as complex-valued functions on $A \times M$ and $C \times M$. For $p \in M$, define $\lambda_p: A \to C$, $\delta_p: \underset{\sim}{C} \to C$ as follows:

$$\lambda_p(a) = \lambda(a, p) \qquad (1.2)$$

$$\delta_p(\Delta) = \delta(\Delta, p) \qquad (1.3)$$

for $p \in M$, $\Delta \in \underset{\sim}{C}$, $a \in A$. (C denotes the complex numbers). Notice that, for fixed p,

$$\lambda_p(a_1 a_2) = \lambda_p(a_1)\lambda_p(a_2)$$
for $a_1, a_2 \in A$ \qquad (1.4)

$$\delta_p(\Delta_1 + \Delta_2) = \delta_p(\Delta_1) + \delta_p(\Delta_2)$$
$$\delta_p(\Delta_1 \Delta_2) = \delta_p(\Delta_1)\delta_p(\Delta_2) \qquad (1.5)$$

for $\Delta_1, \Delta_2 \in \underset{\sim}{C}$.

Let A^d and $\underset{\sim}{C}{}^d$ denote the space of mappings: $A \to C$ and $\underset{\sim}{C} \to C$ satisfying (1.4) and (1.5), respectively. A^d forms an abelian group, called the *dual group* of A, since two elements can be multiplied. Similarly, $\underset{\sim}{C}{}^d$ forms a commutative, associative algebra, since two elements can be added, and multiplied. (1.4) and (1.5) then define mappings: $M \to A^d$ and $M \to \underset{\sim}{C}{}^d$. Often these mappings serve to characterize the whole situation.

Of course, one must relate the elements ψ_p of $\underset{\sim}{D}$ to the Hilbert space H with which we started. Often, this takes the following form: M is a locally compact space, and dp is a "Radon measure"; roughly, this means that

$$\int_M f(p) dp < \infty$$

when $p \to f(p)$ is a continuous complex-valued function of compact support on M. For such a function, define:

$$\psi_f = \int_M f(p) \psi_p . \qquad (1.5)$$

We will not be precise about the definition of the "integral" on the right hand side of (1.5);

various general notions of "integration of vector-valued functions" are possible. For practical purposes, the following property is crucial:

ψ_f can be identified with an element of $\underset{\sim}{D}$, such that:

$$<\psi_f|\psi> = \int_M f(p)^* <\psi_p|\psi> dp \qquad (1.6)$$

for $\psi \in H$.

Now, the completion of the Hilbert space H under the norm $||\psi|| = <\psi|\psi>^{1/2}$ can be identified with a subspace of $\underset{\sim}{D}$. In fact, by the Riesz representation theorem, -- a basic theorem in Hilbert space theory [8] -- it can be identified with the space of linear functionals on H that are continuous with respect to the topology defined by the norm $||\ ||$.

We can then require that:

a) The ψ_f belong to the completion of H.

b) These ψ_f are dense in the completion, with respect to the topology defined by the norm $||\ ||$.

If these properties are satisfied, the

physicists would say that "one can form wave packets with the plane waves ψ_p." Of course, these concepts are related to the "Spectral Theorem" considered by mathematicians [8], but unfortunately it is usually very awkward to work in the precise framework necessitated by the rigorous form of the Spectral Theorem. Certainly, it is essential that those working in this area who want to use the greater flexibility of the Dirac ideas be aware of the rigorous mathematics that is available, and convince themselves of the plausibility of what they are doing in terms of these foundations. Finally, in most of Lie group representation theory it is reasonable and consistent to use the formalism in a semi-heuristic manner, since the needed technicalities can usually be provided by using another formalism, the differential and integral geometry of vector bundles.

After these general remarks, we turn to the study of G as the Poincaré group, which is most important for "partial wave analysis."

2. IRREDUCIBLE REPRESENTATIONS OF SEMIDIRECT PRODUCTS AND THE POINCARÉ GROUP

Continue with the general framework sketched in Section 1. Suppose that A is an an *invariant* abelian subgroup of G, with L another subgroup such that:

$$G = L \cdot A. \qquad (2.1)$$

(2.1) means that every element $g \in G$ can be written in one and only one way as:

$$g = \ell a, \qquad (2.2)$$

with $\ell \in L$, $a \in A$.
G is said to be a *semidirect product* of L and A.

It is surprising how frequently one encounters this sort of group, as a unifying mathematical theme running through many diverse branches of physics. For example, there are very striking analogies between elementary particle and solid-state physics from this point of view. In the former discipline, L = Lorentz group, SO(3,1), A is R^4, the group of translations in four dimensional Euclidean space. In the latter, G is the group mapping a "lattice" in R^3 into itself. A is usually a discrete group, L a "point group", i.e. a finite subgroup of SO(3,R) that has a representation by 3 × 3 matrices with *integer* coefficients. In elementary particle physics, the irreducible representations of G are "elementary particles"; in solid-state physics they represent "elementary excitations." Also, many of the "little subgroups" of the point groups L (whose representations one must find to find all representations of G) are themselves semi-direct products. Luckily, semi-direct product groups form one of the simplest general class of groups from the point of view of classifying representations. The "semisimple groups," for example, are, except for the lowest dimensional cases, SL(2,R) and SL(2,C), much more difficult.

Let us turn to studying a representation D(G) of such a semidirect product group. Let M be a "parameter space" for the eigenvalues of D(A), as explained in Section 1. For p ε M, let λ_p be the element of the dual group, A^d, determined by p. Let $\underset{\sim}{D}^p$ be the subspace of ψ ε $\underset{\sim}{D}$ which are

SCATTERING AMPLITUDE

eigenvectors of $D(A)$ with eigenvalues λ_p, i.e.

$$D(a)(\psi) = \lambda_p(a)\psi \tag{2.3}$$

for $a \in A$

Let us calculate the effect of acting on A by Ad L:

For $\ell \in L$, $a \in A$, $\psi \in \underset{\sim}{D}^p$,

$$D(\ell\ a\ \ell^{-1})(\psi)$$
$$= D(\ell)D(a)D(\ell^{-1})\psi$$
$$= \lambda_p(\ell\ a\ \ell^{-1})\psi,$$

or

$$D(a)(D(\ell^{-1})\psi)) = \lambda_p(\ell\ a\ \ell^{-1})(D(\ell^{-1})\psi)) \tag{2.4}$$

Suppose that L acts as a transformation group on M, in such a way that the mapping $M \to A^d$ intertwines the action of L on both spaces, i.e.

$$\lambda_{\ell^{-1}p}(a) = \lambda_p(\ell\ a\ \ell^{-1}) \tag{2.5}$$

For example, we could define this action if

the mapping $p \to \lambda_p$ of $M \to A^d$ were one-one, hence M were identified with a subspace of A^d. (2.4) then says that this subspace is invariant under the action of AdL on A^d, defined in the following way:

$$\ell(\lambda)(a) = \lambda(\ell^{-1} a \ell) \qquad (2.6)$$

For the moment, we will leave open the more general possibility.

Using (2.5), (2.4) can be rewritten in the following form:

$$D(\ell)(\underset{\sim}{D}^p) \subset \underset{\sim}{D}^{\ell p} \qquad (2.7)$$

for $\ell \in L$, $p \in M$.

This general property of the family of subspaces $\{\underset{\sim}{D}^p : p \in M\}$ is associated with what Mackey calls a "system of imprimitivity". As a variant of his ideas, let us adopt the following definition:

DEFINITION. Let L be a group, $L \to D(L)$ a representation of L by operators on a vector space $\underset{\sim}{D}$. Suppose M is a space on which L acts as a transformation group, and $p \to \underset{\sim}{D}^p$ is a mapping of M into

SCATTERING AMPLITUDE   151

the space of linear subspaces of $\underset{\sim}{D}$. L acts, via D(L), on the linear subspaces of $\underset{\sim}{D}$. We say that the assignment $p \to \underset{\sim}{D}^p$ is a *system of imprimitivity* if it intertwines the action of L on both spaces, i.e. if (2.7) is satisfied.

Now, a "system of imprimitivity" is obviously just a special case of the "vector bundle" idea [5], since the collection $\{\underset{\sim}{D}^p : p \to M\}$, when suitably made into a topological space, forms a vector bundle with M as its base space. Let us denote this bundle by E. A cross-section $\underset{\sim}{\psi} \in \Gamma(E)$ is then a map $p \to \psi(p) \in \underset{\sim}{D}^p$.

What is the relation between $\Gamma(E)$, $\underset{\sim}{D}$ and H? To see this, define a map $\phi: \Gamma(E) \to \underset{\sim}{D}$ as follows:

$$\underset{\sim}{\psi} \to \phi(\psi) = \int_M \underset{\sim}{\psi}(p) dp \qquad (2.8)$$

We now determine the intertwining properties of this map relative to the action of G = LA. First, define G as a transformation group on E as follows:

$$\ell a(\psi) = \lambda_p(a)\ell(\psi) = D(\ell a)(\psi) \qquad (2.9)$$

for $a \in A$, $p \in M$, $\underset{\sim}{\psi} \in \underset{\sim}{D}^p$

This defines E as a "homogeneous vector bundle" of G[5], hence also a representation D' of G by operators $\Gamma(E)$:

$$D'(\ell a)(\underset{\sim}{\psi})(p) = \ell a(\underset{\sim}{\psi}(a^{-1}\ell^{-1}p))$$
$$= \ell a(\underset{\sim}{\psi}(\ell^{-1}p))$$
$$= D(\ell a)(\underset{\sim}{\psi}(\ell^{-1}p)) \qquad (2.10)$$

For $\ell \in L$, $a \in A$, $\psi \in \Gamma(E)$.

THEOREM 2.1. The map ϕ: $\Gamma(E) \to \underset{\sim}{D}$, given by (2.8), is an intertwining map for the representations D' and D of G, providing the measure dp on M is invariant under the action of L.

Proof. For $\ell \in L$, $a \in A$, $\underset{\sim}{\psi} \in \Gamma(E)$

$$\phi D'(\ell a)(\underset{\sim}{\psi}) = \int_M D'(\ell a)(\underset{\sim}{\psi})(p) dp$$

=, using (2.10),

$$\int D(\ell a)\underset{\sim}{\psi}(\ell^{-1}p)dp$$

$$= D(\ell a) \int \underset{\sim}{\psi}(p)dp, \text{ providing dp}$$

is invariant under translation by L,

$$= D(\ell a)\phi(\underset{\sim}{\psi}) \qquad \text{Q.E.D.}$$

This intertwining map is the basic object in Mackey's theory of representations of semidirect products [7]. We will not go into the functional analysis detail needed to make all this rigorous.

In terms of E, we can recognize "irreducible" representations of G as those satisfying the following conditions:

    a) L acts transitively on M, with the map $M \to A^d$ one-one.

    b) Let p_o be a fixed point at M, K the (2.11) isotropy subgroup of L at p_o: (The "little subgroup" of L at p_o.) Then, $D(K)$ maps $\underset{\sim}{D}^{p_o}$ into itself. This "little group" representation should be irreducible.

Then, an irreducible representation of G = LA is determined by a subgroup K of L and an irreducible representation of K. K is not an arbitrary subgroup of L, however. In fact, since the map: $M \to A^d$ is an intertwining operator for the action of L, K must be the isotropy subgroup of L, acting on A^d, at the point $\lambda_{p_o} \in A^d$. We see then that, in the language of induced representations [5, 7], the representation D' is constructed in the following way: Let B be the subgroup KA of G. Then, a representation σ of B on $\underset{\sim}{D}^{p_o}$ can be defined as follows:

$$\sigma(ka) = \lambda_{p_o}(a)D(k) \qquad (2.12)$$

We see that D' is the representation of G "induced" from the representation σ of B.

In case conditions (2.11) are not satisfied, the decomposition of D(G) into irreducible representations is related to the following decomposition operations:

    a) The decomposition of M into orbits of L.
    b) The decomposition of D(K) acting in $\underset{\sim}{D}^{p_o}$ into irreducible representations.
    c) In case the map $M \to A^d$ is not one-one, i.e. K is only a subgroup of the isotropy subgroup of L at λ_{p_o}, the decomposition of representations induced from B to G into irreducible representations. In turn, this depends essentially on the "Plancherel formula" for L and the

SCATTERING AMPLITUDE

Frobenius reciprocity theorem [5].

One encounters these complications in two situations, in the physics of elementary particles:

a) H is the tensor product of several "irreducible" representations.

b) H is the space of solutions of a single-particle wave equation, like the Dirac equation or the Majorana equation.

Finally, we can indicate how all this specializes to the most important case for elementary particle physics: G = Poincaré group, L = Lorentz subgroup, A = translation subgroup. Let $\underset{\sim}{A}$ be the Lie algebra of A, identified with a four-dimensional real vector space. The map Exp: $\underset{\sim}{A} \to A$ is then an isomorphism. Let $\underset{\sim}{a}_1 \cdot \underset{\sim}{a}_2$ denote the Lorentz inner product on $\underset{\sim}{A}$. It is, of course, invariant under Ad L, since Ad L is the defining representation of L = SO(3,1). Using the inner product, $\underset{\sim}{A}^d$ is isomorphic with $\underset{\sim}{A}$, and this isomorphism intertwines the actions of L. This enables us to use the Dirac notations, as the physicists do:

$$\psi \in \underset{\sim}{D}^p \quad \text{if}$$

$$D(\exp(\underset{\sim}{a}))|\psi\rangle = e^{i\underset{\sim}{\lambda} \cdot \underset{\sim}{a}}|\psi\rangle, \qquad (2.13)$$

where $\underset{\sim}{\lambda} \in \underset{\sim}{A}$ is the element corresponding under these isomorphisms, to $\lambda_p \in A^d$. The process of "relaxation of notations" can be carried one step further -- identify $\underset{\sim}{p}$ with $\underset{\sim}{\lambda}$, i.e. consider p as a vector in $\underset{\sim}{A} = R^4$; denote ψ as follows:

$$|\underset{\sim}{p},\ldots\rangle ,$$

where ... denote additional "quantum numbers"

$$(\text{needed if dim } D^{\underset{\sim}{p}} > 1);$$

and rewrite (2.13) as follows:

$$D(\exp(\underset{\sim}{a}))|\underset{\sim}{p} \ldots\rangle = e^{i\underset{\sim}{p}\cdot\underset{\sim}{a}}|\underset{\sim}{p} \ldots\rangle \qquad (2.14)$$

There is even a further convention needed to account for the physicists notations, since they have the habit of introducing $i = \sqrt{-1}$ into formulas by writing one-parameter groups as $\exp(i\ t\ \underset{\sim}{a})$, but we will not attempt to introduce this into our notations.

3. PARTIAL WAVE ANALYSIS OF THE SCATTERING AMPLITUDE

Continue with the same group G = LA as in

SCATTERING AMPLITUDE   157

Section 2. Let H_1; $D_1(G)$; H_2; $D_2(G)$; H_3; $D_3(G)$; H_4; $D_4(G)$ be four representations of G, with parameter spaces M_1, M_2, M_3, M_4. Suppose these representations have the property that the maps $m_j \to A^d$ are one-one, $j = 1, 2, 3, 4$, so that we can label the subspaces $D_j^{\underline{p}_j}$ of D_j by vectors $\underline{p}_j \in R^4 = \underline{A}$. Suppose that:

$$\underline{p}_j \cdot \underline{p}_j = m_j^2, \qquad (3.1)$$

and this number m_j is the same for any point of m_j (m_j is the "mass" of the representation, i.e. of the "particle" depicted by the representation. This number can be identified with the "mass" of the representation determined by the relevant Casimir operator of the Poincaré group [5]). As in (2.14), let $|\underline{p}_j \ldots>$ denote a vector in $D_j^{\underline{p}_j}$, for $\underline{p} \in m_j$.

Our aim is to study maps S: $H_1 \otimes H_2 \to H_3 \otimes H_4$ that intertwine the action of G on these spaces.

Physically, the vectors of $H_{in} = H_1 \otimes H_2$ and $H_3 \otimes H_4 = H_{out}$ represent two particles coming in and out of a collision experiment: If $|in>$ is a vector of $\underline{D}_{in}$ representing two particles coming in towards a collision, $S|in\rangle = |out\rangle$ is the vector of $\underset{\sim}{D}_{out}$ representing the particles coming out of the collision. (Of course, the consideration of only two particles is not sacred -- it is just the simplest situation that can be discussed, both theoretically and experimentally). Since one does not know precisely what happens in collision experiments, say in a particle accelerator, physicists have been attracted to the idea of theories that deal directly with the operator S, particularly since many of the properties of S can be described directly by experiment. However, in the last few years attention has been focussed on less ambitious questions of the relation between certain "kinematic" properties of S, related to the "spin" of particles, certain asymptotic behavior of S at "high energy", the so-called "Regge asymptotic behavior," and relation with experimental phenomenology. We can refer to the book by Eden [3] for a first approximation to this work.

The first step in the analysis of S is to express invariance under L and A in a convenient form. Adopt the following notation for a typical element of $\underset{\sim}{D}_j$, $j = 1, 2, 3, 4$:

$$|\underset{\sim}{p}_j, \alpha_j>, \qquad (3.2)$$

where $\underset{\sim}{p}_j$ is a four-vector such that $\underset{\sim}{p}_j \cdot \underset{\sim}{p}_j = m_j^2$, α_j is a index ranging over the values:

$$1 \leq \alpha_j \leq 2s_j + 1.$$

s_j, is the "spin" of the j-th particle. As α_j varies, the elements (3.2) range over a basis of $D_{\underset{\sim}{j}}^{\underset{\sim}{p}_j}$.

Now, a typical element $|\underset{\sim}{p}_1, \alpha_1> \otimes |\underset{\sim}{p}_2, \alpha_2>$ can be denoted by $|\underset{\sim}{p}_1, \underset{\sim}{p}_2, \alpha_1, \alpha_2>$. Similarly, denote by $|\underset{\sim}{p}_3, \underset{\sim}{p}_4, \alpha_3, \alpha_4>$ an "outgoing state". Now, by general principles, the "matrix elements"

$$S(\underset{\sim}{p}_1, \underset{\sim}{p}_2, \underset{\sim}{p}_3, \underset{\sim}{p}_4; \alpha_1, \alpha_2, \alpha_3, \alpha_4)$$
$$= <\underset{\sim}{p}_3, \underset{\sim}{p}_4, \alpha_3, \alpha_4|S|\underset{\sim}{p}_1, \underset{\sim}{p}_2; \alpha_1, \alpha_2> \qquad (3.3)$$

are "generalized functions" of vectors $\underset{\sim}{p}_1, \underset{\sim}{p}_2, \underset{\sim}{p}_3, \underset{\sim}{p}_4$, i.e. a "generalized function" on R^{16}. (See Gel'fand and Shilov [4] or Chapter VII for an introduction to this concept). To see this, let us suppose that $S: H_1 \otimes H_2 \to H_3 \otimes H_4$ can be extended to a map:

$\underset{\sim}{D}_1 \otimes \underset{\sim}{D}_2 \rightarrow \underset{\sim}{D}_3 \otimes \underset{\sim}{D}_4$. (Of course, this must be assumed to even make sense of the right hand side of (3.3)). Then, if $f(\underset{\sim}{p}_1, \underset{\sim}{p}_2, \underset{\sim}{p}_3, \underset{\sim}{p}_4)$ is a "test function" of the indicated variables, chosen from a suitably defined class (say, "compact support"), the "inner product" $<S|f>$ between S, represented by the left hand side of (3.3), and f is:

$$<S|f> = \int f(\underset{\sim}{p}_1, \underset{\sim}{p}_2, \underset{\sim}{p}_3, \underset{\sim}{p}_4)<\underset{\sim}{p}_3, p_4,$$

$$\ldots |S|\underset{\sim}{p}_1, p_2, \ldots > dp_1\ dp_2\ dp_3\ dp_4 \qquad (3.4)$$

To interpret (3.4), regard "dp_j" as the Lorentz invariant volume element on the "mass-shell" submanifold $p_j^2 = m^2$ of R^4. The integral must then be interpreted as an interated integral. For example, set:

$$f_{\underset{\sim}{p}_3,\ \underset{\sim}{p}_4}(\underset{\sim}{p}_1, \underset{\sim}{p}_2) = f(\underset{\sim}{p}_1, \underset{\sim}{p}_2, \underset{\sim}{p}_3, \underset{\sim}{p}_4).$$

Then,

$$<S|f> = \iint_{M_3 \times M_4} <\underset{\sim}{p}_3, \underset{\sim}{p}_4, \ldots |S| \iint_{M_1 \times M_2}$$

$$f_{\underset{\sim}{p}_3\underset{\sim}{p}_4}(p_1, p_2)\underset{\sim}{p}_1, \underset{\sim}{p}_2, \ldots\ dp_1\ dp_2\ dp_3\ dp_4$$

SCATTERING AMPLITUDE

There are no doubt many interesting concepts in the theory of generalized functions involved in making this sort of analysis completely rigorous; we shall not go into it here.

Now, the invariance of S under translations D(A) implies that the matrix elements, indicated on the left hand side of (3.3), that we will denote by $S(p, \alpha)$, are not C^∞ functions of the p's. The simplest assumption as to their form seems to be the following:

$$S(p, \alpha) = <\text{out}|I|\text{in}>$$
$$+ \delta(p_1 + p_2 - p_3 - p_4) T(p, \alpha), \quad (3.5)$$

(I = identity map)

where T is a C^∞ function of the indicated variables, that is called the *scattering amplitude*. Here, $\delta(p_1 + p_2 - p_3 - p_4)$ denotes the generalized "delta-function" on R^{16} defined as in Chapter VII.

Having expressed translation invariance of S in the form (3.5), we can examine invariance under L. Let us first make precise what happens as L acts on "single particle states". Return to the notation H for a single representation of G, with a state of D^{p_1} denoted by: $|p_1, \alpha_1\rangle$. Then, for $\ell \in L$, and for fixed $\underline{p}_1$ (with $\underline{p}_1$ continued to be identified with an element of R^4), $D(\ell)|\underline{p}_1, \alpha_1\rangle$ is a linear combination of the states $|\ell p_1, \beta_1\rangle$. As $\underline{p}_1$ varies, the coefficients become functions of $\underline{p}_1$. Then, we have relations of the form:

$$D(\ell)|\underline{p}_1, \alpha_1\rangle = \ell(\underline{p})_{\beta_1 \alpha_1} |\ell \underline{p}_1, \beta_1\rangle. \qquad (3.6)$$

Then, $(\ell(\underline{p}_1))_{\alpha_1 \beta_1} = \underline{\ell}(\underline{p}_1)$ are $(2s+1) \times (2s+1)$. matrix valued functions of $\underline{p}_1$, (they contain all the information about the "system of imprimitivity" defining the representation. Of course, $\underline{p}_1$ is restricted by the condition: $\underline{p}_1 \cdot \underline{p}_1 = m^2$).

The intertwining property of S implies that:

$$\langle \ell \text{ out}|S|\ell \text{ in}\rangle = \langle \text{out}|S|\text{in}\rangle$$

for $\ell \in L$. $\qquad (3.7)$

Now, using (3.6)

SCATTERING AMPLITUDE 163

$$\ell|in\rangle = \ell|\underset{\sim}{p}_1, \alpha_1\rangle \otimes \ell|\underset{\sim}{p}_2, \alpha_2\rangle$$

$$= \ell(\underset{\sim}{p}_1)_{\beta_1\alpha_1} \ell(\underset{\sim}{p}_2)_{\beta_2\alpha_2} |\ell\underset{\sim}{p}_1, \ell\underset{\sim}{p}_2, \beta_1, \alpha_1\rangle$$

Hence (3.7), i.e. Lorentz invariance, implies that:

$$S(\underset{\sim}{p}, \underset{\sim}{\alpha}) = \ell(\underset{\sim}{p}_3)^*_{\beta_3\alpha_3} \ell(\underset{\sim}{p}_4)^*_{\beta_4\alpha_4} \ell(\underset{\sim}{p}_1)_{\beta_1\alpha_1}$$

$$\ell(\underset{\sim}{p}_2)_{\beta_2\alpha_2} S(\ell\underset{\sim}{p}, \beta) \qquad (3.8)$$

Combine (3.5) with (3.8):

$$\delta(\underset{\sim}{p}_1 + \underset{\sim}{p}_2 - \underset{\sim}{p}_3 - \underset{\sim}{p}_4) T(\underset{\sim}{p}, \alpha)$$

$$= \delta(\ell\underset{\sim}{p}_1 + \ell p_2 - \ell\underset{\sim}{p}_3 - \ell\underset{\sim}{p}_4) \ell(\underset{\sim}{p}_3)^*_{\beta_3\alpha_3}$$

$$\ell(\underset{\sim}{p}_4)^*_{\beta_4\alpha_3} \ell(\underset{\sim}{p}_1)_{\beta_1\alpha_1} \ell(\underset{\sim}{p}_2)_{\beta_2\alpha_2} T(\ell\underset{\sim}{p}, \beta)$$

Now, $\delta(\ell\underset{\sim}{p}_1 + \ell\underset{\sim}{p}_2 - \ell\underset{\sim}{p}_3 - \ell\underset{\sim}{p}_4) = \delta(\underset{\sim}{p}_1, \underset{\sim}{p}_2, \underset{\sim}{p}_3, \underset{\sim}{p}_4)$, hence (3.8) expresses itself in the form:

$$0 = \delta(\underset{\sim}{p}_1 + \underset{\sim}{p}_2 - \underset{\sim}{p}_3 - \underset{\sim}{p}_4)(T(\underset{\sim}{p}, \alpha) - \ell(\underset{\sim}{p}_3)^*_{\beta_3\alpha_3}$$

$$\ell(\underset{\sim}{p}_4)^*_{\beta_4\alpha_4} \ell(\underset{\sim}{p}_1)_{\beta_1\alpha_1} \ell(\underset{\sim}{p}_2)_{\beta_2\alpha_2} T(\ell\underset{\sim}{p}, \beta)),$$

whence the condition:

$$T(\underline{p}, \alpha) = \ell(\underline{p}_3)^*_{\beta_3\alpha_3} \ell(\underline{p}_4)^*_{\beta_4\alpha_4}$$

$$\ell(\underline{p}_1)_{\beta_1\alpha_1} \ell(\underline{p}_2)_{\beta_2\alpha_2} T(\underline{p}, \beta) \qquad (3.9)$$

when $\underline{p}_1 + \underline{p}_2 + \underline{p}_3 + \underline{p}_4 = 0$.

We can interpret (3.9) geometrically in the following way: Let N be the submanifold of $R^{16} = R^4 \times R^4 \times R^4 \times R^4$ given by the condition:

$$\underline{p}_1 + \underline{p}_2 + \underline{p}_3 + \underline{p}_4 = 0 \qquad (3.10)$$

$$\underline{p}_1^2 = m_1^2, \quad \underline{p}_2^2 = m_2^2, \quad \underline{p}_3^2 = m_3^2, \quad \underline{p}_4^2 = m_4^2$$

L acts as a transformation group on this space. The set of functions $T(\underline{p}, \alpha)$ transform, as indicated by (3.9), under transformation by L.

Then, the value of the T's at one point of N determines the values at all points of the orbit of L of that point. Let us then examine the structure of the orbits of L on N. Introduce the following functions on N:

SCATTERING AMPLITUDE

$$s = (p_1 + p_2)^2$$
$$t = (p_1 - p_3)^2. \qquad (3.11)$$

s is the "center of mass energy", t the "momentum transfer". s and t are constant on the orbits of L. Since

$$p_j^2 = m_j^2,$$

we have:

$$s = m_1^2 + m_2^2 + 2 p_1 \cdot p_2$$
$$t = m_1^2 + m_3^2 - 2 p_1 \cdot p_3 \qquad (3.12)$$

Conversely, if two points of N have the same value of s and t, there is Lorentz transformation carrying one into the other. To see this, suppose $p = (p_1, p_2, p_3, p_4)$, $p' = (p_1', p_2', p_3', p_4')$ are these two points. It follows from (3.12)

$$p_1 \cdot p_2 = p_1' \cdot p_2', \; p_1 \cdot p_3 = p_1' \cdot p_3'.$$

Now, square the relation (3.10) to deduce that:

$$\underline{p}_2 \cdot \underline{p}_3 = \underline{p}_2' \cdot \underline{p}_3'.$$

This implies that the corresponding inner products between elements of p and p' are the same -- it is readily shown that this implies there is a Lorentz transformation transforming one into the other.

Thus, we see that the map: $N \to R^2$ obtained by assigning (s, t) to $\underline{p}$, via formula (3.11), passes to the quotient to define an isomorphism of the orbit space L/N with R^2. We refer to [6], Chapter 25 for a discussion of what is meant by "orbit space." (The treatment given there only covers the case where N is a Riemannian manifold, and L is a group of isometrics. If the m_j^2 are positive, j = 1, 2, 3, 4, one can, in fact, show that these conditions are satisfied. In any case, many of the ideas can be generalized to the non-Riemannian case.)

Recall the following terminology. (It is the general terminology adapted to this situation.) A point of N is called a *principal point* relative to N if the isotropy subgroup of L at this point is the identity. A point is called a *maximal point* if the isotropy subgroup at the point is finite. A point of N is called a *singular point* if the isotropy subgroup at this point is of non-zero dimension.

Suppose we assume for simplicity $m_j^2 > 0$, so that we are in the Riemannian case. Then, N_1^o, the set of principal points, is an open, dense, subset of N, and $L \backslash N^o$, its orbit space, is a manifold. (In fact, one can show that N_o consists of the points $\underline{p} = (\underline{p}_1, \underline{p}_2, \underline{p}_3, \underline{p}_4)$ such that the matrix $(\underline{p}_i \cdot \underline{p}_j)$ has rank 3). By the general theory described in [6], N^o can be written as a product $(L \cdot \underline{p}^o) \times (L \backslash N^o)$, where $\underline{p}^o = (\underline{p}_1^o, \ldots, \underline{p}_4^o)$ is a fixed point of N^o. Now (s, t) are coordinates for the orbit space $L \backslash N^o$.

Thus, the functions $T(\underline{p}, \alpha)$ can be written in N^o as functions of point of L and variables (s, t), since L parameterizes the orbit space $L \backslash N^o$. This gives a representation of the scattering amplitude T in terms of the "invariants" of the action of L, N -- namely s and t -- and the functions $\ell(\underline{p}_j^o)_{\alpha_j \beta_j}$ $j = 1, 2, 3, 4$, considered as functions of $\ell \in L$. However, this representation may be expected to break down and/or develop singularities as one approaches the boundary of N^o in N. This leads to what physicists call *kinematic singularities* [1].

This representation of T in terms of invariants (s, t) and function $\ell(p_j{}^o)_{\alpha_j \beta_j}$ on L is the starting point for studies of the analyticity properties of T, [3], e.g. the Mendelstam representation. There are many details in this theory that are but special cases of general facts about transformation groups and their singularities.

Now, we turn to the description of the partial wave expansion. Let $\underline{p} = (\underline{p}_1, \underline{p}_2, \underline{p}_3, \underline{p}_4)$ be a point of N. Let $K(\underline{p})$ be the subgroup of L consisting of the transformation that leave $\underline{p}_1 + \underline{p}_2$ invariant. (If we are in the "physical region", with $s = (\underline{p}_1 + \underline{p}_2)^2 > 0$, then $K(\underline{p})$ is isomorphic to $SO(3,R)$.) Now, given $\underline{p} \in N$, define a mapping $\pi_{\underline{p}}$: $K(p) \to N$ by the following formula:

$$\pi_{\underline{p}}(k) = (\underline{p}_1, \underline{p}_2, k\underline{p}_3, k\underline{p}_4) \qquad (3.13)$$

for $k \in K(\underline{p})$, $\underline{p} = (\underline{p}_1, \underline{p}_2, \underline{p}_3, \underline{p}_4)$.

(Notice that the right hand side of (3.13) belongs to N if $\underset{\sim}{p}$ does, so that the formula really has the range that is claimed for it.) Notice that $K(\underset{\sim}{p})$ is really a "kinematic symmetry." The scattering amplitude is not invariant under it, but, since it leaves invariant the "kinematics", the amplitude can be expanded in terms of matrix elements of representations of $K(\underset{\sim}{p})$. This, roughly, is what is meant by "partial wave analysis."

Explicitly, consider the following functions in $K(\underset{\sim}{p})$, with $\underset{\sim}{p}$ considered as being held fixed:

$$T_{\underset{\sim}{p},\,\alpha_1\alpha_2\alpha_3\alpha_4}(\ell) = \ell^{-1}_{\beta_3\alpha_3}(\ell\underset{\sim}{p}_3)$$
$$\ell^{-1}_{\beta_4\alpha_4}(\underset{\sim}{p}_4) T(\underset{\sim}{p},\,\alpha_1\alpha_2\beta_3\beta_4). \qquad (3.14)$$

Then, for $\ell \in K(p)$ we can propose possible candidates for scattering amplitudes by giving $T_{\underset{\sim}{p},\alpha}(K(\underset{\sim}{p}))$ as a sum -- whether continuous or discrete, -- over matrix elements of representations of the group $K(\underset{\sim}{p})$, as explained in Chapter II. (The coefficients in these expansions are then functions of $\underset{\sim}{p}$, which, in fact, depend only on s and certain spin indices.) If one chooses only the unitary representations of $K(\underset{\sim}{p})$ to use in this expansion, this leads to the usual "partial wave expansion" in terms of spherical harmonics with "discrete indices". (Of course, this also requires that $K(\underset{\sim}{p})$ is compact, which it will be when $\underset{\sim}{p}$ is in the "physical region.") However, there is also the possibility of using matrix elements of non-unitary representations of $K(\underset{\sim}{p})$ -- these are the "Regge expansions" of the scattering amplitude. The representations of $K(\underset{\sim}{p})$ that contribute *discretely* to this expansion are the "Regge poles." Obviously, there will be a very delicate interplay between the "analytic" and "asymptotic" properties of T and the properties of the coefficients in the various partial wave expansions -- these interconnections have been one of the major objects of study in elementary particle physics in the last ten years.

There is another more "geometric" interpretation of the partial wave analysis that is also of interest -- particularly since it generalizes very unambiguously to situations where there are a greater number of particles.

SCATTERING AMPLITUDE   171

For $\underset{\sim}{p} \in N$, let $N(\underset{\sim}{p})$ be the set of $\underset{\sim}{p}' = (\underset{\sim}{p}_1', \underset{\sim}{p}_2', \underset{\sim}{p}_3', \underset{\sim}{p}_4') \in N$ such that:

$$\underset{\sim}{p}_1' = \underset{\sim}{p}_1, \quad \underset{\sim}{p}_2' = \underset{\sim}{p}_2$$

$$\underset{\sim}{p}_3' + \underset{\sim}{p}_4' = \underset{\sim}{p}_1 + \underset{\sim}{p}_2.$$

Let us calculate $N(\underset{\sim}{p})$ more explicitly, in case we are in the "physical region", i.e. the $m_j^2 = p_j^2$ are positive, and the zeroth components of these four vectors are positive. One then shows (using the Schwarz inequality) that $s = (p_1 + p_2)^2$ is also positive. We can then -- without essential loss in generality -- suppose that:

$$p_1 + p_2 = (m, 0, 0, 0), \text{ with } m = m_1 + m_2$$

Suppose then that:

$$p_3' = (E_3, \vec{p}_3), \quad p_4' = (E_4, \vec{p}_4)$$

Then, the conditions determining $N(\underset{\sim}{p})$ are:

$$E_3^2 - \vec{p}_3^2 = m_3^2$$

$$E_4^2 - \vec{p}_4^2 = m_4^2$$

$$E_3 + E_4 = m, \quad E_3 > 0, \quad E_4 > 0$$

$$\vec{p}_3 + \vec{p}_4 = 0$$

Eliminate E_4 and p_4, and the conditions become:

$$E_3^2 = m_3^2 + \vec{p}_3^2$$

$$(m - E_3)^2 - \vec{p}_3^2 = m_4^2,$$

or:

$$(m - \sqrt{m_3^2 + \vec{p}_3^2})^2 = \vec{p}_3^2 + m_4^2$$

or

$$m^2 - 2m\sqrt{m_3^2 + \vec{p}_3^2} + m_3^2 = m_4^2,$$

or

$$2m\sqrt{m_3^2 + \vec{p}_3^2} = m^2 + m_3^2 - m_4^2, \quad \text{or}$$

$$\vec{p}_3^2 = \left(\frac{m^2 + m_3^2 - m_4^2}{2m}\right)^2 - m_3^2. \qquad (3.15)$$

Now, one readily verifies that the right hand side of (3.15) is positive -- hence $N(p)$ is, as a differentiable manifold, a sphere.

The basic idea in the "geometric" approach to "partial wave analysis" is to regard the expansion as defining an expansion of the scattering amplitudes T as an expansion in terms of eigenfunctions of certain "geometrically defined" differential operators on the manifolds $N(\underset{\sim}{p})$. For example, if the particles are all spin zero, i.e. there are no indices α, the appropriate operator is the Laplace -- Beltrami operator [6] of the Riemannian metric on $N(\underset{\sim}{p})$ induced from the given imbedding: $N(\underset{\sim}{p}) \subset R^8 = R^4 \times R^4$. (Of course, with R^4 given the Minkowski metric). If there are spin-indices, the appropriate choice is a "bundle Laplacian", defined by putting a metric on the vector bundle whose fibres are represented by the indices α. (The point is that one wants the resulting operator to be invariant under the action of the group $K(\underset{\sim}{p})$ on $N(\underset{\sim}{p})$ and those vector bundle whose fibres are determined by the indices α).

This geometric point of view is useful also for the analysis of collision experiments with a greater number of particles. Suppose, for example, that one considers two particles in, and *three* out. Then, we are given $\underline{p}$ of the form:

$$(\underline{p}_1, \underline{p}_2, \underline{p}_3, \underline{p}_4, \underline{p}_5).$$

Let $N(\underline{p})$ be the space of such vectors $\underline{p}'$ such that:

$$\underline{p}_1 = \underline{p}_1'; \quad \underline{p}_2 = \underline{p}_2'.$$

$$\underline{p}_3' + \underline{p}_4' + \underline{p}_5' = \underline{p}_1 + \underline{p}_2.$$

$$\underline{p}'_j{}^2 = m_j{}^2, \quad j = 1, 2, 3, 4, 5.$$

To determine $N(\underline{p})$ more explicitly, separate the 4-vectors p' into zeroth and vector components.

$$\underline{p}_j' = (E_j, \vec{p}_j), \quad j = 3, 4, 5.$$

In addition, we can suppose without essential loss in generality that:

$$\underline{p}_1' + \underline{p}_2' = (m, 0, 0, 0),$$

with $m = m_1 + m_2$.

SCATTERING AMPLITUDE

Then, a) $E_3 + E_4 + E_5 = cm$

b) $\vec{p}_3 + \vec{p}_4 + \vec{p}_5 = 0$

c) $E_3^2 - \vec{p}_3^2 = m_3^2$ (3.16)

d) $E_4^2 - \vec{p}_4^2 = m_4^2$

e) $E_5^2 - \vec{p}_5^2 = m_5^2$.

Combining, a), b) and e),

$$(m - E_3 - E_4)^2 - (\vec{p}_3 + \vec{p}_4)^2 = m_5^2, \quad \text{or}$$

$$(m^2 + E_3^2 + E_4^2 + 2E_3E_4) - 2m(E_3 + E_4)$$
$$+ \vec{p}_3^2 + 2\vec{p}_3 \cdot \vec{p}_4 + \vec{p}_4^2 = m_5^2,$$

or

$$m^2 + m_3^2 + m_4^2 + 2E_3E_4 - 2m(E_3 + E_4)$$
$$+ 2\vec{p}_3 \cdot \vec{p}_4 = m_5^2 \quad (3.17)$$

Thus, on eliminating E_3 and E_4 from (3.17), using (3.16) d), we obtain a single equation for the vector $(\vec{p}_3, \vec{p}_4)$ in R^6. In fact, one can show that the hypersurface thus obtained in R^6 is diffeomorphic to a five-dimensional sphere. See Dragt

[2] for an analysis of the situation. A remarkable fact is that the group SU(3) acts transitively on this space, hence plays the same role as the "little group" $K(\underset{\sim}{p})$ in the 2-body situation. Presumably, expansion of the scattering amplitude in terms of representations of SU(3) and/or eigenfunctions of bundle-Laplacians on homogeneous vector bundles on this space is very interesting physically.

BIBLIOGRAPHY

1. K. Bardacki and G. Segré, Some Conspiracy and Superconvergence Properties of Scattering Amplitudes in the Helicity Formalism, Nuovo Cimento, 53A, 56-68 (1968).

2. A. J. Dragt, Classification of Three Particle States According to SU(3); Relativistic Three Particle SU(3) States, J. of Math. Phys., 6, 533-553; 1621-1625 (1965).

3. R. J. Eden, High Energy Collisions of Elementary Particles, Cambridge University Press, 1967.

4. I. M. Gel'fand and G. E. Shilov, Generalized Functions, Academic Press, New York, 1964.

5. R. Hermann, Lie Groups for Physicists, W. A. Benjamin, New York, 1966.

6. R. Hermann, Differential Geometry and the Calculus of Variations, Academic Press, New York, 1968.

7. G. W. Mackey, Induced Representations, W. A. Benjamin, New York, 1969.

8. F. Riesz and B. S. Nagy, Functional Analysis, Ungar, New York, 1955.

CHAPTER VI

PARTIAL WAVE ANALYSIS AS A PROBLEM
IN GROUP REPRESENTATION THEORY

1. INTRODUCTION

A recent tendency in elementary particle physics is to replace the "explicit", classical treatment of partial wave analysis to be found in most quantum mechanics books by a more "abstract" analysis using group-theoretic principles. For example, we can cite work by Jacob and Wick [3], Joos [4], Moussa and Stora [14], Toller [11, 12] and Roffman [10]. A basic mathematical idea in this approach is to study the scattering operator as an intertwining operator between two Hilbert spaces (the "incoming" and "outgoing" states) in which the Poincaré group acts as a unitary group of operators.

Now, the theory of intertwining operators has been extensively developed in the mathematical literature - most notably in the works of G. W. Mackey [6, 7]. Our aim in this chapter is to present material that bridges the gap between these two theories.

2. A GENERAL ALGEBRAIC VIEWPOINT IN FUNCTIONAL ANALYSIS SUGGESTED BY S-MATRIX THEORY

In general, one may remark that the standard theory of Hilbert spaces, Banach spaces, etc., is not well-suited to the needs of quantum physics (or group-representation theory either, for that matter). On the other hand, the theory of topological vector spaces is too general. Gel'fand and coworkers have developed an intermediate theory ("rigged Hilbert spaces" [1]) that is, no doubt, the best available. However, even this theory is not ideal; after all, it was developed mainly for the sake of applications to the theory of partial differential equations, and inevitably different features must be emphasized when applications to other disciplines are considered. The ideal theory probably would be one that combined features of the Gel'fand theory and that of van Neumann's "Rings of operators" [9] (or the version of it developed by Mackey [6, 7] for applications to group-representation theory). In this paper, we will describe in imprecise terms some features such a theory might have: Hopefully, such a sketch might stimulate the experts in functional analysis to develop the needed concepts in rigorous terms.

We will be dealing with a class $\underset{\sim}{H}$ of complex vector spaces. Individual elements of $\underset{\sim}{H}$ will be denoted by H, H', etc. We will also be dealing with a class $\underset{\sim}{A}$ of linear transformations with range and domain included in the vector-space elements of $\underset{\sim}{H}$. Typical elements of $\underset{\sim}{A}$ will be denoted by A, A', B, etc. We shall suppose that the composition (or operator-product) of two elements of $\underset{\sim}{A}$ again has an $\underset{\sim}{A}$ (providing the composition is defined, of course, i.e. if the domains and ranges match up), and that the sum of two elements of $\underset{\sim}{A}$ with the same domain and range also lies in $\underset{\sim}{A}$.

We shall also suppose that the elements of $\underset{\sim}{A}$ have defined on them an adjoint operator, defined by *. If A is an operator: $H \to H'$, A^* is an operator: $H' \to H$, and it satisfies the usual algebraic identities for the adjoint of a matrix:

$$(A_1 + A_2)^* = A_1^* + A_2^*$$

$$(A_1 A_2)^* = A_2^* A_1^*$$

$$(cA)^* = c^* A^* \text{ for each complex scalar } c.$$

(c^*, for a complex number, denotes complex conjugate $1^* = 1$ (1 denotes the identity operator).

Finally, we want to consider a function on operators of $\underset{\sim}{A}$ with properties similar to the trace of a square matrix. We will suppose that there is a subclass $\underset{\sim}{A}'$ of $\underset{\sim}{A}$ such that:

    a) Each $A \in \underset{\sim}{A}'$ maps H into H, for some $H \in \underset{\sim}{H}$.

    b) With each $A \in \underset{\sim}{A}'$ is associated a complex number: trace (A), satisfying trace $(A_1 + A_2) =$ trace $A_1 +$ trace A_2, trace $(A_1 A_2) =$ trace $(A_2 A_1)$ for $A_1, A_2 \in \underset{\sim}{A}'$.

Given $H, H' \in \underset{\sim}{H}$, we will denote the elements of $\underset{\sim}{A}$ which map H into H' by:

$\underset{\sim}{A}(H, H')$

Thus, defining:

$$\langle A_1 | A_2 \rangle = \text{trace}\,(A_1^* A_2) \tag{2.1}$$

for $A_1, A_2 \in \underset{\sim}{A}(H, H')$,

(whenever, of course, the right hand side of (2.1) is defined, i.e. for $A_1^* A_2 \in \underset{\sim}{A}'$) makes $\underset{\sim}{A}(H, H')$ into something like a Hilbert space.

No doubt these ideas could be formulated precisely in the language of categories and functors, but we will not be so formal; we will keep as close as possible to the physicist's pragmatic attitude towards mathematical formalisms.

Let G be a group. Suppose that for each $H \in \underset{\sim}{H}$, a representation $g \to D(g)$ of G by operators on H is given, and that $D(g)$, for each $g \in G$, belongs to $\underset{\sim}{A}(H, H)$. We shall suppose that each of these representations is *unitary* in the sense that

$$D(g)^* = D(g^{-1}) \quad \text{for} \quad g \in G.$$

An operator $A \in \underset{\sim}{A}(H, H')$ is called on *intertwining operator* if:

$$D'(g)A = AD(g) \quad \text{for} \quad g \in G.$$

Let $I(H, H')$ denote the vector space of these intertwining operators.

As an example of such a mathematical structure suggested by physics, consider $H, H' \in \underset{\sim}{H}$, and regard the vectors of H and H' as the "incoming" and "outgoing" states of a physical system before and after scattering, with the representations $D(G)$, $D'(G)$ on H and H' regarded as "symmetries" of the physical system. The "scattering operator" S should be considered as an intertwining operator: $H \to H'$, i.e. as an element of $I(H, H')$. Such intertwining operators can be decomposed into integrals of "elementary" ones which map a single subspace of H that is irreducible under $D(G)$ into one that is irreducible under $D'(G)$. Crudely, this is the "partial wave decomposition" of S. Let us formulate it more precisely as follows.

Suppose Λ is a space which parametrizes a set of irreducible unitary representations of G, i.e. for $\lambda \in G$, $g \to D_\lambda(g)$ is an irreducible representation of G by unitary operators on a space $H_\lambda \in \underset{\sim}{H}$. Let M be a space, and let $\pi: M \to \Lambda$ be a

PROBLEM IN GROUP REPRESENTATION   185

mapping of M onto Λ. For $\lambda \in \Lambda$, let M_λ be the fibre of π, i.e. the set of $p \in M$, with $\pi(p) = \lambda$.

For each $p \in M_\lambda$, let $A_p \in I(H, H_\lambda)$, $A_p' \in I(H', H_\lambda)$ be given operators. Then, the operator

$$A_p'^* A_p$$

is an "elementary" intertwining operator: $H \to H'$. We suppose decomposition of the form:

$$S = \int_M f(p) A_p'^* A_p d_p, \qquad (2.2)$$

where d_p is a measure on M. This is the "partial wave decomposition" of S. We suppose the normalization of the A_p', A_p chosen so that the following "Plancherel formula" holds.

$$\langle S | S' \rangle = \int_M f(p)^* f'(p) dp$$

for S' of the form

$$\int_M f'(p) A_p'^* A_p d_p.$$

Symbolically, then,

$$\langle A_p'^* A_p | A_{p'}'^* A_{p'} \rangle = \delta(p, p').$$

Also, symbolically,

$$f(p) = \langle A_p'^* A_p | S \rangle$$

EXAMPLE. Suppose G is a compact group (e.g. G = SU(2)), with its irreducible unitary representations labelled by a discrete index ℓ. Suppose further that ℓ_1, ℓ_2, ℓ_3, ℓ_4 are four such indices.

$$H = H_{\ell_1} \times H_{\ell_2}, \quad H' = H_{\ell_3} \times H_{\ell_4}.$$

Then, D(G) acting on H is $D_{\ell_1} \times D_{\ell_2}$, D'(G) acting in H' is $D_{\ell_3} \times D_{\ell_4}$. Suppose also that: In the decomposition of D(G) and D'(G) into irreducible representations, each representation of G occurs at most once. Then, we can take M as the set of indices (ℓ) itself. The spaces $I(H, H_\ell)$, $I(H', H_\ell)$ are at most one-dimensional.

Let us use the Dirac notations to describe an orthonormal basis for H_ℓ, namely $|\ell, m\rangle$, where m is an integer index running over a certain range (depending on ℓ, of course). Then, we can define the generalization of the Wigner 3-j symbols as the matrix elements of the operators B_ℓ:

PROBLEM IN GROUP REPRESENTATION   187

$$\begin{pmatrix} \ell_1 & \ell_2 & \ell \\ m_1 & m_2 & m \end{pmatrix} = \pm \langle \ell m | B_\ell | \ell_1 m_1 \times \ell_2 m_2 \rangle$$

(We will not attempt here to keep the sign conventions straight.)

We also clearly have:

$$\langle B_\ell'^{*} B_\ell | B_{\ell'}'^{*} B_{\ell'} \rangle = 0 \quad \text{if} \quad \ell \neq \ell'.$$

B_ℓ itself is normalized so that $\langle B_\ell | B_\ell \rangle = 1$. Thus, A_ℓ can be chosen as follows

$$A_\ell = \frac{B_\ell}{\langle B_\ell'^{*} B_\ell | B_\ell'^{*} B_\ell \rangle^{1/4}}$$

(2.2) then gives a decomposition of S in terms of the 3-j symbols. As Joos, Sertorio and Toller have remarked [4, 11, 12] a similar construction for the case G = Poincaré group yields the partial-wave decomposition of the scattering amplitude, thus giving an elegant group-theoretic description of this decomposition.

Let us now ask whether these ideas can be rephrased even more sharply in our categorical-group theoretic language. Let H, H' be two spaces in H, with representation $g \to D(g)$, $D'(g)$ of G by unitary operators in H and H'. Let $\underset{\sim}{A}(H, H')$ be the vector space of all operators in $\underset{\sim}{A}$ that map H into H'. $G \times G$ acts on $\underset{\sim}{A}(H, H')$, via a representation that we shall label $D' \times D$.

For $A \in \underset{\sim}{A}(H, H'), (g_1, g_2) \in G \times G$,

$$(D' \times D)(g_1, g_2) = D'(g_1)AD(g_2^{-1}) \qquad (2.3)$$

Notice that $I(H, H')$ consists of the elements of $\underset{\sim}{A}(H, H')$ that are invariant under the diagonal subgroup of $G \times G$. One may suspect therefore that the decomposition (2.2) may be obtained from the decomposition of $(D' \times D)(G \times G)$ into irreducible representations, then expanding an $S \in I(H, H')$ in terms of this decomposition.

To lend support to this idea, let us make a few further remarks. Let $H'' \in \underset{\sim}{H}$, with $D''(G)$ an *irreducible* representation of G on H''. Consider a fixed $A_0 \in I(H, H'')$, and $B_0 \in I(H'', H')$. Construct the subspace of $\underset{\sim}{A}(H, H')$ consisting of all operators of the form:

$$B_0 A'' A_0,$$

with $A'' \in \underset{\sim}{A}(H'', H'')$. We can denote this by:

$$B_0 \underset{\sim}{A}(H'', H'') A_0$$

Then, obviously:

$$(D' \times D)(G \times G)(B_0 \underset{\sim}{A}(H'') A_0) = B_0 \underset{\sim}{A}(H'') A_0,$$

i.e. this subspace of $\underset{\sim}{A}(H, H')$ is invariant and (at least in the finite dimensional case) irreducible, and in fact equivalent to the action of $D'' \times D'$ on $\underset{\sim}{A}(H'', H'')$. Presumably, in the finite dimensional case also, $\underset{\sim}{A}(H', H')$ admits such a decomposition into irreducible pieces (at least if G is a type-1 group). Notice that the elements of $B_0 \underset{\sim}{A}(H'', H'') A_0$ that are invariant under the diagonal subgroup of $G \times G$ are those of the form: constant $\times B_0 A_0$, so that the decomposition of an $S \in I(H, H')$ into these irreducible pieces would, in fact, just be a decomposition of type (2.2). In turn, the decomposition of the representation $D \times D'$ of $G \times G$ on $\underset{\sim}{A}(H, H')$ into irreducible components involves the Fourier analysis of functions on $G \times G$. (We shall review this in the next section.) This then establishes, at least in principle, the connection between Fourier analysis on groups and partial wave analysis.

3. THE IRREDUCIBLE PROJECTION OPERATORS DEFINED BY INTEGRATION OVER THE GROUP

"Partial wave analysis" in the form used by physicists involves not the rather abstract algebraic formalism described in Section 2, but a more explicit definition of the decomposition of intertwining operators into irreducible ones by integration over the group. Now, it is well-known how to do this for compact groups. We will reformulate it for this case in terms of the formalism sketched in Section 2 -- our formalism has the virtue of suggesting immediately the needed generalization to the case of non-compact G.

Suppose then that $\underset{\sim}{H}, \underset{\sim}{A}$, and G are as in Section 2. However, suppose that G is compact. We suppose given a unitary representation $g \to D(g)$ of G by operators in H, with H a given element in $\underset{\sim}{H}$. Suppose, in fact, that H is a Hilbert space, and the adjoint operation is defined as usual for Hilbert spaces. Let Λ be a space whose points λ

label equivalence classes of irreducible, unitary representations of G, denoted by D_λ, on a Hilbert space H_λ. We, of course, suppose that each H_λ and each $D_\lambda(g)$ belong to our basic "category" of spaces and operators.

Of course, since G is compact, Λ is a discrete space -- however, we choose our notation as if Λ were a more general sort of measure space. Let $d\lambda$ denote a "measure" on Λ. Formulate the Fourier decomposition of a function $g \to f(g)$ on G as follows:

$$\hat{f}(\lambda) = \int_G f(g) D_\lambda(g^{-1}) dg. \qquad (3.1)$$

(dg is a two-sided invariant volume element on G)

$$f(g) = \int_\Lambda \text{trace}(\hat{f}(\lambda) D_\lambda(g)) c(\lambda) d\lambda \qquad (3.2)$$

$c(\lambda)$ is a function on Λ, the "Plancherel measure." Of course, in the case G is compact, Λ is discrete, with elements $\lambda_1, \lambda_2, \ldots$, (3.2) takes the well-known form:

$$f(g) = \sum_n \text{trace}(\hat{f}(\lambda_n) D_{\lambda_n}(g)) c(\lambda_n), \qquad (3.3)$$

with

$$c(\lambda_n) = \dim H_{\lambda_n} \qquad (3.4)$$

We refer to (Chapter 2) for a short discussion of these matters.

Let us formulate the decomposition of $D(G)$ into irreducible representations in the following manner. Suppose that M is a space, with $\pi: M \to \Lambda$ a mapping of M onto Λ. Denote a typical point of M by p, a measure on M by dp. Suppose given, for each $p \in M$, an intertwining map:

$$B_p: H \to H_{\pi(p)} \qquad (3.5)$$

Then,

$$B_p^* B_p \text{ is an intertwining map: } H \to H.$$

The image $B_p^* B_p$ is then an intertwining map of H onto a subspace of H in which $D(G)$ acts irreducibly. Suppose that:

$$\Psi = \int_M B_p^* B_p(\Psi) dp \quad \text{for} \quad \Psi \in H \qquad (3.6)$$

Further, suppose that the subspaces $B_p^* B_p(H)$ are

"orthogonal" as p runs over M, i.e.

$$B_p^* B_p B_{p'}^* B_{p'} = 0 \quad \text{for} \quad p \neq p'. \tag{3.7}$$

Then, if A is an operator on $\underset{\sim}{A}(H, H)$,

$$\text{trace } A = \int_M \text{trace}(B_p A B_p^*) dp. \tag{3.8}$$

(3.8) is, of course, equivalent to (3.6)-(3.7), in the case G is compact, and, say, that M is discrete. (To see this, one has only to take A to be the projection of H onto the one-dimensional subspace spanned by Ψ.) The point is that in the case of non-compact G, (3.8), phrased solely in terms of our "category," makes more sense. Thus, for $g \in G$,

$$\text{trace}(AD(g)) = \int_M \text{trace}(B_p AD(g) B_p^*) dp. \tag{3.9}$$

On the other hand, apply (3.2) to the function

$$f_A(g) = \text{trace}(AD(g)):$$

$$f_A(g) = \int_\Lambda \text{trace}(\hat{f}_A(\lambda) D_\lambda(g)) c(\lambda) d\lambda \tag{3.10}$$

Let us suppose that the integral integral over M can be decomposed into an integral over Λ and the fibres of π in the following way:

$$\int_M h(p)dp = \int_\Lambda (\int_{\pi'(\lambda)} h(p)d_\lambda p)c(\lambda)d\lambda \qquad (3.11)$$

where, for each $\lambda \in \Lambda$, $d_\lambda p$ is a volume element over the fibre $\pi^{-1}(\lambda)$ of the map π "above" the point λ. Let us apply (3.11) to (3.9):

$$f_A(g) = \int_\Lambda (\int_{\pi^{-1}(\lambda)} \text{trace}(B_p AD(g) B_p^*)d_\lambda p)c(\lambda)d\lambda$$

since B_p is an intertwining operator,

$$\int_\Lambda (\int_{\pi^{-1}(\lambda)} \text{trace}(B_p AB_p^* D_{\pi(p)}(g))d_\lambda p)c(\lambda)d\lambda$$

$$= \int_\Lambda (\int_{\pi^{-1}(\lambda)} \text{trace}(B_p AB_p^* d_\lambda p)D_\lambda(g)c(\lambda)d\lambda. \qquad (3.12)$$

Then, equating (3.12) with (3.10), we have:

$$0 = \int_\Lambda \text{trace}(\int_{\pi^{-1}(\lambda)} (B_p AB_p^*)d\,p - \hat{f}_A(\lambda))D_\lambda(g)c(\lambda)d\lambda$$

This says that the inverse Fourier transform of the operator-valued function:

$$\lambda \to \hat{f}_A(\lambda) - \int_{\pi^{-1}(\lambda)} B_p AB_p^* d_\lambda p$$

must be zero. Accordingly, the function itself

PROBLEM IN GROUP REPRESENTATION 195

must be zero, i.e.

$$\int_{\pi^{-1}(\lambda)} (B_p A B_p^*) d_\lambda p = \int_G \text{trace}(AD(g)) D_\lambda(g^{-1}) dg. \quad (3.13)$$

This is the basic formula: The left-hand side of (3.13) determines the projection of H onto the vectors that transform under D(G) like the representation $D_\lambda(G)$, while the right hand side exhibits this projection as an integral over the group.

Now, we have exhibited the basic projection (3.1) in a form that one can see makes sense (at least) for non-compact Lie groups G as well as compact ones. Notice, however, that its derivation would require that the function

$$g \to f_A(g) = \text{trace}(AD(g)) \quad (3.14)$$

be "rapidly decreasing" an G, e.g. square-integrable. For example, this will happen, for the case where A is the projection of H onto a one-dimensional subspace, only if the matrix elements of D(g) are square integrable, i.e. if the representation D(G) is a direct sum of "discrete series." However, we can proceed as follows, using a generalized-function point of view.

Consider a class $\underset{\sim}{A}' \subset \underset{\sim}{A}(H)$ of "test operators," i.e. ones for which the function (3.14) belongs a certain class of rapidly-decreasing test-functions on G. Suppose that a given operator $A \in \underset{\sim}{A}(H)$ defines a linear functional on $\underset{\sim}{A}'$, i.e.

$$<A|A'> \equiv \text{trace}(A^*A') \qquad (3.15)$$

is defined and finite for $A' \in \underset{\sim}{A}$. $A' \to B_p A' B_p^*$ defines a linear mapping

$$\underset{\sim}{A}'(H) \to \underset{\sim}{A}(H_{\pi(p)}).$$

It can be extended to $\underset{\sim}{A}$ by saying that an element $A'' \in \underset{\sim}{A}(H_{\pi(p)})$ is its image if:

$$<A''|B_p A' B_p^*> = <A|A'> \qquad (3.16)$$

for all $A' \in \underset{\sim}{A}'$.

For example, applied to the case: G = Poincaré group, this would give a precise, group-theoretic way of describing the growth-properties of the scattering amplitude, or, more precisely, defining the scattering amplitude as a "generalized function," without following the physicist's usual method of exhibiting the scattering amplitude as a function of the Lorentz invariants. Aside from a possible gain of mathematical elegance, this group-theoretic method of describing the properties of the scattering amplitude without using the Lorentz invariants might avoid some of the well-known diseases of "kinematic singularities." (A simpler example to consider: When describing the analytic properties of a function on a sphere that has prescribed transformation properties under rotations, one can work in a classical way by introducing spherical coordinates. This is analogous to the introduction of the Lorentz invariant in describing the scattering amplitude. However, the singularities of spherical coordinates complicate the description of the properties of the function, and they usually take a more elegant and understandable form when described in a coordinate-free, group-theoretical way.)

4. CROSSING SYMMETRY IN A GROUP-THEORETIC FRAMEWORK

Another ingredient in the melange of ideas called "S-matrix theory" is that of "crossing symmetry." In this section, we will discuss the group-theoretic foundation of this concept -- our treatment is well-known (and, in fact, has been given in the same form by Toller [12]) but we will go over it again, in order to make it available in the language we are developing in these notes.

Let us consider the case a scattering operator, with two particles coming in, two going out. (The generalization to more particles is straightforward.) Consider then a group G, four Hilbert spaces H_1, H_2, H_3, H_4, and unitary representations

$$g \to D_i(g), \quad i = 1, 2, 3, 4$$

of G by operators on these spaces. Set:

$$H = H_1 \otimes H_2, \quad H' = H_3 \otimes H_4.$$

To define the "crossing" operation, suppose that $\bar{H}_1$, $\bar{H}_3$ are also Hilbert spaces, carrying unitary representations denoted by $\bar{D}_1$, $\bar{D}_3$ of G, with $\alpha_1: H \to \bar{H}_1$, $\alpha_3: H_3 \to \bar{H}_3$ anti-unitary mappings that intertwine the given representations of G. Set:

$$\bar{H} = \bar{H}_3 \otimes H_2, \quad \bar{H}_1 \otimes H_4 = \bar{H}'$$

Define a mapping α: $\underset{\sim}{A}(H, H') \to \underset{\sim}{A}(\bar{H}, \bar{H}')$ as follows: For $S \in \underset{\sim}{A}(H, H')$, $\alpha(S)$ is the element such that

$$\langle \bar{\Psi}_1 \times \Psi_4 | \alpha(S) | \bar{\Psi}_3 \times \Psi_2 \rangle$$
$$= \langle \alpha_3^{-1}(\bar{\Psi}_3) \times \Psi_4 | S | \alpha_1^{-1}(\bar{\Psi}_1) \times \Psi_2 \rangle \qquad (4.1)$$

One readily sees that $\alpha(S)$ is really well-defined by this formula. Also,

$$\alpha(I(H, H')) \subset I(\bar{H}, \bar{H}'). \qquad (4.2)$$

Of course, $\bar{D}_1$, $\bar{D}_3$ are "unphysical" representations, if G is the Poincaré group. To see this, suppose, for example, that

$$(P_\mu), \quad 0 \leq \mu \leq 3,$$

are the generators of the translation subgroup of the Lie algebra of the Poincaré group. Suppose that $|p_1\rangle$ are the "plane-wave" states of H_1, i.e.

$$D_1(P_\mu)|p_1\rangle = i(p_1)_\mu |p_1\rangle \qquad (4.3)$$

(p_1 is a 4-vector, $(p_1)_\mu$ denotes its components). Then,

$$D_1(P_\mu)\alpha_1|p_1\rangle = \alpha_1 D_1(P_\mu)|p_1\rangle$$

$$= \alpha_1(i(p_1)_\mu|p_1\rangle)$$

$$= -i(p_1)_\mu|p_1\rangle. \qquad (4.4)$$

Then, as far as states are concerned, the effect of α_1 (in a spinless situation) is the transformation

$$p_1 \to -p_1.$$

Let us calculate $\alpha(S)$ in terms of scattering amplitudes. Suppose that D_1, D_2, D_3, D_4 are irreducible, spinless representations of the connected Poincaré group G, of masses m_1, m_2, m_3, m_4. Denote the plane-wave states of H_1, H_2, H_3, H_4 by $|p_1\rangle$, $|p_2\rangle$, $|p_3\rangle$, $|p_4\rangle$. Suppose $S \in I(H, H')$:

$$\langle p_3, p_4|S-1|p_1, p_2\rangle$$

$$= \delta(p_1 + p_2 - p_3 - p_4)f_S(p_1, p_2, p_3, p_4). \qquad (4.5)$$

Of course, (4.4) expresses translation invariance of S: Lorentz invariance is expressed by

$$f_S(\ell p_1, \ell p_2, \ell p_3, \ell p_4) = f_S(p_1, p_2, p_3, p_4) \qquad (4.6)$$

PROBLEM IN GROUP REPRESENTATION 201

with ℓ an element of the connected Lorentz group.

Similarly,

$$\langle \bar{p}_1, p_4 | \alpha(S) | \bar{p}_3, p_2 \rangle$$
$$= \delta(\bar{p}_3 + p_2 - \bar{p}_1 - p_4) f_{\alpha(S)}(\bar{p}_3, p_2, \bar{p}_1, p_4) \quad (4.7)$$

By (4.1), the left hand side of (4.7) is

$$\langle \alpha_3^{-1}(\bar{p}_3), p_4 | S | \alpha_1^{-1}(\bar{p}_1), p_2 \rangle$$
$$= \langle -\bar{p}_3, p_4 | S | -\bar{p}_1, p_2 \rangle$$
$$= \delta(p_2 - \bar{p}_1 + \bar{p}_3 - p_4) f_S(-\bar{p}_1, p_2, -\bar{p}_3, p_4),$$

or

$$f_{\alpha(S)}(\bar{p}_3, p_2, \bar{p}_1, p_4) = f_S(-\bar{p}_1, p_2, -\bar{p}_3, p_4) \quad (4.8)$$

Now, (4.8) can define $f_{\alpha(S)}$ as a possible "physical" scattering amplitude for a scattering operator

$$H_3 \otimes H_2 \to H_1 \otimes H_4$$

providing f_S can be analytically continued away from the region where it is defined via formula

(4.5), namely the subset of R^{16} defined by the following conditions:

$$p_1 + p_2 - p_3 - p_4 = 0$$

$$p_1^2 = m_1^2; \; p_2^2 = m_2^2; \; p_3^2 = m_3^2; \; p_4^2 = m_4^2$$

$$(p_1)_0 > 0; \; (p_2)_0 > 0; \; (p_3)_0 > 0; \; (p_4)_0 > 0$$

(4.9)

(m_1, m_2, m_3, m_4 are the masses of the irreducible unitary representations D_1, D_2, D_3, D_4 of the connected Poincaré group G). Of course, this is just the usual way of stating "crossing symmetry" in S-matrix theory.

We now ask: Suppose, indeed, that f_S is the restriction to the subset (4.9) of a holomorphic function of these variables *made complex*. Is there an equivalent way to state this condition in purely group-theoretic language?

The following way seems to be the most natural:

Realize H_1 explicitly as the set of all complex-valued functions $p_1 \to \Psi(p_1)$ of a *real* 4-vector p_1, with

$$p_1^2 = m_1^2, \quad (p_1)_0 > 0. \tag{4.10}$$

Now, let H_1^c be the vector space of all (C^∞) functions $p_1 \to \Psi(p_1)$ of the *complex* four-vector p_1, with

$$p_1^2 = m_1^2.$$

The restriction of such a function to the real subset (4.10) defines a linear mapping: $H_1^c \to H_1$. Now, the representation D_1 of the real Poincaré group G by operators on H_1 can be extended in the obvious way to a representation on H_1^c. This representation of G can now be extended to a representation D_1^c of G^c, the "complexification" of G (which is the semi-direct product of the complex Lorentz group SO(4,C) and its 4-dimensional complex-vector representation.

A similar construction can be performed for H_2, H_3, H_4, leading to representations D_2^c, D_3^c, D_4^c. $H_1 \otimes H_2$ can now be realized on the complex-valued functions $(p_1, p_2) \to \Psi(p_1, p_2)$ of *real* four vectors (p_1, p_2), satisfying:

a) $p_1^2 = m_1^2$; $p_2^2 = m_2^2$;

b) $(p_1)_0 > 0$, $(p_2)_0 > 0$. (4.11)

Similarly, $H_3 \otimes H_4$ is defined as a function space.

$H_1^c \otimes H_2^c$ is now realized as the space of all C^∞ complex-valued functions $\Psi(p_1, p_2)$ of *complex* four-vectors restricted by (4.11a) -- $H_3^c \otimes H_4^c$ is defined similarly.

Consider an operator S^c: $H_1^c \otimes H_2^c \to H_3^c \otimes H_4^c$ that intertwines the action of G^c on both spaces. It can be written, in at least a formal way, as:

$$S^c(\Psi)(p_3, p_4) = \int \delta(p_1 + p_2 - p_3 - p_4)$$
$$f^c(p_1, p_2, p_3, p_4)\Psi(p_1, p_2)dp_1 dp_2, \quad (4.12)$$

Here, $dp_1 dp_2$ means a volume element on the subset (4.11a) of complex-four vectors that is invariant under complex Lorentz transformations. We can define "analyticity" of the operator by requiring that S^c map arbitrary functions into holomorphic (= complex analytic) functions and by requiring that the adjoint operator of S^c map arbitrary functions into anti-holomorphic functions. (An anti-holomorphic function is one whose complex-conjugate is holomorphic.) If one wants to prescribe a certain singularity set for the function f^c, one can do this by requiring that S^c only be defined on functions in $H_1^c \otimes H_2^c$ that are zero in the neighborhood of this subset.

This procedure generalizes readily to arbitrary spin representations of the Poincaré-group, using the vector-bundle method for defining these representations. [2] Thus we have succeeded in our aim of describing in a more group-theoretic way (in terms of representations of the complex Poincaré group) several of the key ideas of S-matrix theory, e.g. "analyticity" of the scattering amplitude and the associated idea of crossing-symmetry. Of course, this point of view has recently been described by E. Roffman [10] also, but it will be convenient for later work to present these ideas in this language.

BIBLIOGRAPHY

1. I. M. Gel'fand and N. Ya. Vilenkin, Generalized Functions, IV, Academic Press, New York, 1964.

2. R. Hermann, Lie Groups for Physicists, W. A. Benjamin, New York, 1966.

3. M. Jacob and G. C. Wick, On the General Theory of Collisions for Particles with Spin, Ann. Phys. $\underline{7}$, 404-428 (1959).

4. H. Joos, Complex Angular Momentum and the Representations of the Poincaré Group with Space-Like Momentum, in Lectures in Theoretical Physics, 1964, University of Colorado Press, Boulder, Colorado, 1964.

5. H. Jost, The General Theory of Quantized Fields, American Math. Society, Providence, R.I., 1965.

    D. Kastler, Introduction a l'electrodynamique quantique, Dunod, Paris, 1961.

    A. Kurosh, The Theory of Groups, Chelsea Pub. Co., New York, 1955.

6. G. W. Mackey, The Theory of Group Representations, Lecture notes, Univ. of Chicago, 1955.

7. G. W. Mackey, Induced Representations, W. A. Benjamin, Inc., New York, 1968.

8. P. Moussa and R. Stora, Some Remarks on the Product of Irreducible Representations of the Inhomogeneous Lorentz Group, in Lectures in Theoretical Physics, 1964, University of Colorado Press, Boulder, Colorado, 1965.

9. M. A. Naimark, Normed Rings, P. Noordhoff, Groningen, 1964.

10. E. Roffman, Complex Inhomogeneous Lorentz Groups and Complex Angular Momentum, Phys. Rev. Letters $\underline{16}$, 210 (1966).

11. L. Sertorio and M. Toller, Nuovo Cimento $\underline{33}$, 413 (1964).

12. M. Toller, On the Group-Theoretical Approach to Complex Angular Momentum and Signature, CERN preprint, 1967.

CHAPTER VII

REMARKS ON THE USE OF TRANSFORMATION GROUPS IN QUANTUM FIELD THEORY

1. CONNECTIONS BETWEEN VECTOR BUNDLES AND QUANTUM FIELDS

Quantum field theory has two sides that are of great mathematical interest. On the one hand, there are the "hard" analytical questions of the actual existence of the fields as definite mathematical objects, the analytical properties of the various operators, the existence of scattering states, etc. These are the deepest questions, of course, and accordingly most of the work of mathematical physicists in recent years has been directed towards them. However, there is another class of problems that have received less emphasis from mathematicians, but is no less important scientifically because of its relation to the more "successful" side of elementary particle physics, the study of symmetries. In this approach, one assumes that there is some sort of answer to the first type of problem, and investigates how the symmetries may be imbedded in the scheme. In the next few sections we will present some work in this direction, emphasizing the connection with modern differential geometry. (For the notations of manifold theory, we refer to [1, 4, 7, 11].

Consider a manifold M on which a Lie group G acts as a transformation group. (All data will be of differentiability class C^∞ unless mentioned otherwise.) Let F(M) denote the ring of C^∞, complex-valued functions on M. A typical point of M will be denoted by p; if g ε G, gp denotes the transform of p by g.

Suppose π: E → M is a vector bundle over M, [5] i.e.:

    a) $\pi(E) = M$

    b) For each p ε M, the fibre $\pi^{-1}(p)$ is a

QUANTUM FIELD THEORY 211

vector space (say, over the complex number).

A *cross-section* of E is typically denoted by γ, is a map: M → E such that:

$$\gamma(p) \; \varepsilon \; \pi^{-1}(p) \quad \text{for all} \quad p \; \varepsilon \; M.$$

Let Γ(E) denote the set of all these cross-sections. They can be added, and multiplied by scalars, hence form a vector space. They also form a module over the ring F(M):

$$\text{For} \quad f \; \varepsilon \; F(M), \; \gamma \; \varepsilon \; \Gamma(E), \; (f\gamma)(p) = f(p)\gamma(p).$$

This module structure will play a strong role when we discuss later the theory of connections in vector bundles, and the "Yang-Mills fields".

Suppose that G also acts on a linear way on the vector bundle E. This means that, for g ε G, p ε M, g maps the fibre $\pi^{-1}(p)$ by a linear isomorphism onto the fibre $\pi^{-1}(gp)$. This linear action of G defines a representation D(G) of G by operators on Γ(E):

$$\text{For} \quad g \; \varepsilon \; G, \; \gamma \; \varepsilon \; \Gamma(E),$$
$$D(g)(\gamma)(p) = g\gamma(g^{-1}p)$$

DEFINITION. Suppose given a subspace Γ' of Γ(E) which is invariant under the operators D(G). A *quantum field associated with Γ' and G* is defined by giving:

  a) A Hilbert space H.

  b) A representation D' of G by unitary operators on H.

  c) A linear mapping φ: Γ' → $\underset{\sim}{A}$(H) which intertwines the action of G on these two spaces. ($\underset{\sim}{A}$(H) denotes the space of operators: H → H. The action of G on $\underset{\sim}{A}$(H) is as before: If A ε $\underset{\sim}{A}$(H), g ε G, the transform of A by g is the operator D'(g)AD'(g^{-1}).)

This is the natural generalization of the classical idea of "quantum field." [14] Here, M is space-time, R^4, G = Poincaré group, with its usual transformation group action on R^4. In this case, use x to denote a point of M. Suppose E is the product M × V of M with a vector space V. The action of G on E is determined by a representation ρ of L (= homogeneous Lorentz group) by operators on V: Since G is the semidirect product of L and the invariant subgroup of translations, ρ can be extended to a representation of G by operators on

V which is the identity on the translation subgroup. Then,

$$g(x, v) = (g(x), \rho(g)(v))$$

for $x \in M$, $v \in V$.

An element $\gamma \in \Gamma(E)$ can be identified with a map, which we also denote by γ, of $M \to V$. The representation $D(G)$ is then defined by:

$$D(g)(\gamma)(x) = \rho(g)(\gamma(g^{-1}x)).$$

Of course, this definition does not attempt to cover the notion of "causality" nor to make precise the analytic properties that one wants the "test functions" in Γ' to have. Our main goal is to emphasize certain algebraic topics, and to develop the relations with group representation theory.

Let $\underset{\sim}{D}(E)$ be the set of differential operators on the cross-sections of Γ. An element, say Δ, is then a linear mapping $\Delta: \Gamma(E) \to \Gamma(E)$ such that, when written out explicitly in terms of local coordinates for M and E, involves a linear differential operator in the classical sense. Δ is said

214   FOURIER ANALYSIS ON GROUPS

to be of *r-th degree* if, when written and in terms of local coordinates, Δ involves only the partial derivatives of the coordinates up to and including the r-th degree. Let $\underset{\sim}{D}^r(E)$ be the set of those r-th degree operators. Then,

$$\underset{\sim}{D}^r(\Gamma)\underset{\sim}{D}^s(\Gamma) \subset \underset{\sim}{D}^{r+s}(\Gamma). \qquad (1.1)$$

(The product on the left hand side of 1.1) is the ordinary operator product.)

$$[\underset{\sim}{D}^r(\Gamma), \underset{\sim}{D}^s(\Gamma)] \subset \underset{\sim}{D}^{r+s-1}(\Gamma) \qquad (1.2)$$

The action of G on the cross-sections of E defines a representation of G, that we also denote by $D(G)$, of G by operators on $\underset{\sim}{D}(E)$:

For $g \in G$, $\gamma \in \Gamma(E)$, $\Delta \in \underset{\sim}{D}(E)$,

$$D(g)(\Delta)(\gamma) = D(g)\Delta D(g^{-1})(\gamma) \qquad (1.3)$$

Suppose that for $\phi: \Gamma' \to \underset{\sim}{A}(H)$ defines a quantum field, as defined above, and that $\Delta \in \underset{\sim}{D}(E)$, with

$$\Delta(\Gamma') \subset \Gamma'$$

Define $\Delta(\phi)$ as a map: $\Gamma' \to \underset{\sim}{A}(H)$ as follows:

$$\Delta(\phi)(\gamma) = \phi(\Delta(\gamma))$$

for $\gamma \in \Gamma'$. (1.7)

Then, $\Delta(\phi)$ intertwines the action of G on Γ' on $\underset{\sim}{A}(H)$ if:

$$D(g)(\Delta) = \Delta \quad \text{for} \quad g \in G \qquad (1.8)$$

i.e. if Δ is invariant under G. In particular, this gives us a way of saying that ϕ satisfies a linear differential equation: $\Delta = 0$.

In particular, if Δ satisfies (1.8), we can construct "free" quantum fields satisfying a linear partial differential equation, $\Delta = 0$, generalizing the classical construction using Fock space [3, 8, 13]. The first step in this construction is to let

$$\Gamma' = \{\gamma \in \Gamma(E); \Delta(\gamma) = 0\}, \qquad (1.9)$$

and to impose a Hermitian symmetric, real-bilinear form on Γ', i.e. to try to define a scalar "inner product"

$$\langle \gamma | \gamma' \rangle \qquad (1.10)$$

Between two elements of Γ' that satisfies all the algebraic laws of the inner product between two vectors in a Hilbert space, but is not necessarily positive definite. Once one has such an inner product, the Fock space construction [3, 8, 13] will give a quantum field.

To define such an inner product, we may proceed as follows. Let us suppose, in fact, that Δ is a first order differential operator, i.e.

$$\Delta \in \underset{\sim}{D}^1(E).$$

Let N be a submanifold of M. N is said to be a *non-characteristic* submanifold for Δ if:

> For each non-zero solution γ of $\Delta\gamma = 0$, the restriction of γ to N is not identically zero. (1.11)

Let γ_N denote the restriction to N of a cross-section of E. It is then a cross-section of the vector bundle E restricted to N, that we may denote by E_N. Thus, if $\Gamma(E_N)$ denotes the space of cross-sections of E_N, $\gamma_N \in \Gamma(E_N)$. We may then hope to define an inner product on Γ' (given by (1.9)) by integrating over N the values of an inner product on the fibres of the vector bundle E_N. However, a peculiarity of this construction is that the inner product must depend on the submanifold N, and cannot be an inner product defined on the fibres of the V once and for all.

One way to do this in a natural differential-geometric manner may be described as follows:

Suppose dim N = n. Then, the tangent space to N at a point p ε N, denoted by N_p, is an n-dimensional subspace of M_p, the tangent spaces to points of M. ($\mathcal{G}^n(M)$ is called the *Grassman bundle* over M.) Denote a typical "point" of $\mathcal{G}^n(M)$ by δ_p, i.e. δ_p is an n-dimensional subspace of M_p. Let us suppose that we are given a mapping which assigns to each δ_p ε $\mathcal{G}^n(M)$ a Hermitian inner product, denoted by $<\ |\ >^{\delta p}$, on the fibre $\pi^{-1}(p)$ of E lying over p. With such a geometric-object given, we may define the inner product (1.10) for elements of Γ' (defined by (1.9)) as follows:

$$<\gamma|\gamma'> = \int_N <\gamma(p)|\gamma'(p)>^{N_p} d_N p \qquad (1.12)$$

Here, $d_N p$ denotes a volume element for the

submanifold N.

(1.12) may be put in a more convenient form as follows: Suppose $(\ ,\ )^p$ is a *fixed* Hermitian form on the fibre $\pi^{-1}(p)$ of E. Then, we may try to define (1.12) in the following form:

$$\langle \gamma(p) | \gamma'(p) \rangle^{N_p}_{\ p} = (\alpha(\gamma(p)), \gamma'(p))^p \qquad (1.13)$$

where α is a linear transformation $\pi^{-1}(p) \to \pi^{-1}(p)$. α, of course, depends on the subspace N_p. Thus, we may regard α as a mapping:

$$\mathcal{G}^n(M) \to \text{(space of linear operators on fibres of E)} \qquad (1.14)$$

Such an object may be regarded as a differential geometric version of what the physicists call a "current". (More precisely, such an object defines a current when the system is "quantized".) Now, one must determine to what extent (1.12)-(5.13) is independent of N, and whether it is invariant under the action of G on the solutions of $\Delta = 0$. This will involve "integrability conditions," and is usually associated with the physical name "conserved current". We will not investigate these conditions at this point.

However, let us illustrate these generalities with the usual case considered by physicists: M is Euclidean space, R^4, with points of M now denoted by 4-vectors

$$x = (x_\mu), \quad 0 \leq \mu \leq 3, \text{ summation convention}$$

Suppose that E is the direct product M × V, where V is a vector space. Let G be the Poincaré group, the semidirect product of the homogeneous Lorentz group L, with the group of translations T. Then, T acts trivially on the fibres of E. L acts via a linear representation ρ of L by operators on V.

$$\rho(\ell)(x, v) = (\ell x, \rho(\ell)(v)) \qquad (1.15)$$

Suppose Δ is of the form:

$$\Delta = \alpha^\mu \frac{\partial}{\partial x_\mu} + \alpha \qquad (1.16)$$

where (α^μ), α is a set of operators on V.

A cross-section γ of E can be identified with a map $x \to \gamma(x)$ of M into V. Let N be a 3-dimensional subspace of M. A point γ of N then determines a vector $y_N(p)$ of R^4, the "normal vector" of the tangent space to submanifold N. (Here, it is most convenient to define the normal vector by using the Lorentz inner product

$$x \cdot x' = g_{\mu\nu} x_\mu x'_\nu,$$

where

$$g_{\mu\nu} = \begin{cases} 0 & \text{if } \mu \neq \nu \\ 1 & \text{if } \mu = \nu = 0 \\ -1 & \text{if } \mu = \nu = 1, 2, \text{ or } 3 \end{cases} \quad (1.17)$$

Let (,) denote a fixed inner product on the vector space V. Then, we may define:

$$\langle \gamma | \gamma' \rangle = \int_N (g_{\mu\nu} y_\mu(p) \alpha^\nu \gamma(p), \gamma'(p)) d_N p. \quad (1.18)$$

One may work out the "conservation" conditions, following the usual pattern.

Having described how these basic ideas of quantum field theory may be extended to arbitrary manifolds and vector bundles, we turn to the elucidation of the Yang-Mills idea [12] in this language.

QUANTUM FIELD THEORY

2. YANG-MILLS FIELDS AND THE DIFFERENTIAL GEOMETRY OF VECTOR BUNDLES

We will describe the notations again. All manifolds, maps, etc. will be of differentiability class C^∞. If M is a manifold, F(M) denotes the ring of complex-valued C^∞ functions on M. If $\pi: E \to M$ is a vector bundle over M, $\Gamma(E)$ denotes the vector space of cross-sections, which is, in fact, also a module over F(M).

Let V(M) denote the set of derivations of F(M), i.e. on $X \in V(M)$ is defined as a linear map: $f \to X(f)$ of V(M) into itself such that:

$$X(f_1 f_2) = X(f_1)f_2 + f_1 X(f_2) \text{ for } f_1, f_2 \in F(M).$$

V(M) is a Lie algebra relative to the Jacobi bracket operation:

$$[X, Y](f) = X(Y(f)) - Y(X(f))$$

$$\text{for } X, Y \in V(M), f \in F(M).$$

V(M) is also an F(M)-module:

$$(fX)(f_1) = fX(f_1) \text{ for } f_1 f_1 \in F(M), X \in V(M).$$

An element X of V(M) is called a *vector field* on M. One can prove that X can be identified with a cross-section of a vector bundle on M, called the *tangent bundle*, and denoted by T(M). The fibre of T(M) above a point p ε M, denoted by M_p, called the tangent space to M at p, and defined as the set of all linear mappings v: F(M) → (complex numbers), such that:

$$v(f_1 f_2) = v(f_1)f_2(p) + f_1(p)v(f_2).$$

DEFINITION. A *linear connection* for E is defined as a C-bilinear map V(M) × Γ(E) → Γ(E), denoted by (X, Ψ) → $\nabla_X \Psi$, such that:

a) $\nabla_{fX}\Psi = f\nabla_X\Psi$ for   f ε F(M), Ψ ε Γ(E)

b) $\nabla_X(f\Psi) = X(f)\Psi + f\nabla_X\Psi$

    for   f ε F(M), Ψ ε Γ(E)   (2.1)

∇ is called the *covariant derivative* operation defining the connection.

Now, the rule (2.1b) prevents ∇ from defining a tensor field. However, the fact that the first

QUANTUM FIELD THEORY 223

term on the right hand side of (2.1b) is independent of ∇ implies that the difference of two such connections is a tensor field. Indeed, let ∇' be another such covariant derivative operation. For $X \in V(M)$, $\Psi \in \Gamma(E)$, define:

$$A(X, \Psi) = \nabla_X \Psi - \nabla'_X \Psi.$$

Then,

$$A(fX, \Psi) = fA(X, \Psi) = A(X, f\Psi)$$

for $f \in F(M)$, $X \in V(M)$, $\Psi \in \Gamma(E)$ \hfill (2.2)

This says that the operation $V(M) \times \Gamma(E) \to \Gamma(E)$ defined by $A(X, \Psi)$ is $F(M)$ linear, which is "well-known" [4, 7] to be equivalent to the tensorial nature of A. In fact, for $p \in M$, one can define the "value" of A at p, denoted by A_p. It is a bilinear map $M_p \times E_p \to E_p$ (M_p = tangent space to M at p), such that

$$A(X, \Psi)(p) = A_p(X(p), \Psi(p))$$

for $p \in M$, $X \in V(M)$, $\Psi \in \Gamma(E)$.

The assignment $p \to A_p(\, , \,)$ defines a cross-section of another vector bundle over M.

In quantum field theory, one encounters the following special case of this notion. $M = R^4$, Minkowski space, with Euclidean coordinates x_μ ($0 \leq \mu \leq 3$) $E = M \times V$, where V is a complex-vector space. Then, an element of $\Gamma(E)$ can be identified with a map $\Psi: M \to V$. Suppose ∇ is the "flat" connection, i.e.

$$\nabla_{(\frac{\partial}{\partial x_\mu})} \Psi = \frac{\partial \Psi}{\partial x}.$$

Set: $A_\mu = A(\frac{\partial}{\partial x_\mu},\ )$. Thus $A_\mu(x) = A_x(\frac{\partial}{\partial x_\mu},\ )$, is a linear map: $V \to V$. For example, if each A_μ is multiplication by a pure-imaginary scalar, $A_\mu(x)$ would represent the potentials of an electromagnetic field.

$$\nabla'_{\frac{\partial}{\partial x_\mu}} \Psi = (\frac{\partial}{\partial x_\mu} - A_\mu)\Psi \qquad (2.3)$$

This is the familiar rule of "minimal electromagnetic coupling." The Yang-Mills generalization consists in allowing the A_μ to lie in a given Lie algebra of linear transformations on V.

Now, one basic reason for the rule (2.3) is its covariant nature relative to gauge transformations. We shall now investigate this for the general vector bundle case.

3. GAUGE TRANSFORMATION OF LINEAR CONNECTIONS; A COHOMOLOGY INTERPRETATION

Return to the case where E is a general linear vector bundle. A diffeomorphism $g: E \to E$ is a *gauge transformation* of E if:

> For $p \in M$, g maps the fibre $\pi^{-1}(p) = E_p$ linearly and isomorphically onto the fibre above another point $gp \in M$. (3.1)

Thus, g is a fibre-preserving, linear diffeomorphism of E into itself. Such a g acts on cross-sections.

For $\Psi \in \Gamma(E)$, define $D(g)(\Psi) \in \Gamma(E)$ as follows:

$$D(g)(\Psi)(p) = g\Psi(g^{-1}p) \qquad (3.2)$$

The gauge transformations form a group, denoted by $G(E)$. The assignment $f \to D(g)$ defines a representation of $G(E)$ by linear operators on $\Gamma(E)$.

Suppose a covariant derivative operator ∇ defines a linear connection for the vector bundle E. Consider a group G of gauge transformations on E. Each $g \in G$ defines a transform $D(g)(\Delta)$ of Δ:

$$D(g)(\nabla)_X \Psi = D(g)(\nabla_{D(g^{-1})X} D(g^{-1})\Psi)$$

for $X \in V(m)$, $\Psi \in \Gamma(E)$. \hfill (3.3)

(Here, $D(g)(X)$ is the transform of the vector field X by the diffeomorphism g defines on M:

$$D(g)(X)(f) = g^{-1*}(Xg^*(f))$$

for $X \in V(M)$, $f \in F(M)$.

$g^*(f)$ is the function $p \to f(gp)$, for $p \in M$. Then, the difference $\nabla - D(g)\nabla$ is a tensor field:

$$\nabla_X \Psi - D(g)(\nabla)_X \Psi = A_g(X, \Psi)$$

for $X \in V(M)$, $\Psi \in \Gamma$

Then, for $g, g_1 \in G$,

QUANTUM FIELD THEORY

$$D(g_1)(A_g(X, \Psi)) = D(g_1)(\nabla_X \Psi) - D(g_1)(D(g)(\nabla)_X \Psi)$$

$$= D(g_1)(\nabla)_{D(g_1)X} D(g_1)(\Psi)$$

$$\quad - D(g_1 g)(\nabla)_{D(g_1)X} D(g_1)\Psi$$

$$= \nabla_{D(g_1)X} D(g_1)\Psi - A_{g_1}(D(g_1)X, D(g_1)\Psi)$$

$$\quad - \nabla_{D(g_1)X} D(g_1)\Psi + A_{g_1 g}(D(g_1)X, D(g_1)\Psi)$$

$$= A_{g_1 g}(D(g_1)X, D(g_1)\Psi) - A_{g_1}(D(g_1)X, D(g_1)\Psi) \quad (3.5)$$

Now, as we have mentioned, each tensor-like field $A_g(X, \Psi)$ is a mass-section of a vector bundle over M, that we will denote by E'. (In fact, E'_p, the fibre over a point p of M, consists of the tensor-product of M_p with the vector space of linear mappings of E_p into itself.) The action of G on E and M also defines a representation that we will denote by D', of G by operators of $\Gamma(E')$:

$$D'(g)(A)(X, \Psi) = D(g)(A(D(g)X, D(g)\Psi)) \quad (3.6)$$

Consider the mapping $g \to A_g(\ ,\ )$ as defining a 1-cochain of G with coefficients in the representation D'. (Here we use the notions of group cohomology. See [6, 9].) Let us denote this 1-cochain by ω. Let us see if (3.5) has an interpretation in terms of group cohomology:

$$d\omega(g_1, g) = \omega(g_1) + D'(g_1)(\omega(g)) - \omega(g_1 g).$$

for $g_1, g \in G$.

Now,

$$D'(g_1)(\theta(g))(X, \omega) = D'(g_1)(A_g)(X, \Psi)$$

$$= D(g_1)(A_g(D(g_1^{-1})X, D(g^{-1})\Psi))$$

$$\theta(g_1)(X, \omega) = A_g(X, \omega)$$

$$\theta(g_1 g)(X, \omega) = A_{g_1 g}(X, \omega).$$

Comparing with (2.5), we see that:

$$d\omega = 0 \qquad (3.7)$$

i.e. ω determines a 1-cocycle of G relative to the representation D'.

Suppose now that ∇' is another linear connection for E. The action of G on it defines

QUANTUM FIELD THEORY 229

another 1-cocycle of G, that we can denote by ω'. Now,

$$\nabla_X \Psi - \nabla_X' \Psi = A(X, \Psi),$$

where $A(\ ,\ )$ is a fixed element of $\Gamma(E')$. This determines a 0-cochain of G relative to D', that we denote by θ. It is readily seen that

$$d\theta = \omega - \omega'.$$

Hence, we have proved:

THEOREM 3.1. G acting on the vector bundle E leaves invariant some linear connection for E if and only if the 1-cocycle of G relative to $D'(G)$ defined by (3.4) cobounds.

4. THE CURVATURE OF A VECTOR-BUNDLE CONNECTION

Suppose E, M, π, $\Gamma(E)$, V(M)G, ∇, are as in Sections 1-3. For X_1, $X_2 \in V(M)$, $\Psi \in \Gamma(E)$ let:

$$\underset{\sim}{F}(X_1, X_2)(\Psi) =$$
$$= \nabla_{X_1}(\nabla_{X_2}\Psi) - \nabla_{X_2}(\nabla_{X_1}\Psi) - \nabla_{[X_1, X_2]}\Psi \qquad (4.1)$$

This is the *curvature tensor* for the connection. (It is readily seen that it is F(M)-linear, which justifies calling it a tensor.)

For example, suppose that ∇ is given by (2.3), as

$$\nabla_{\frac{\partial}{\partial x_\mu}} (\Psi) = \frac{\partial \Psi}{\partial x_\mu} - A_\mu \Psi.$$

Then,

$$\underset{\sim}{F}_{\mu\nu} = \underset{\sim}{F}\left(\frac{\partial}{\partial x_\mu}, \frac{\partial}{\partial x_\nu}\right) = \left(\nabla_{\frac{\partial}{\partial x_\mu}} \nabla_{\frac{\partial}{\partial x_\nu}} - \nabla_{\frac{\partial}{\partial x_\nu}} \nabla_{\frac{\partial}{\partial x_\mu}}\right)(\Psi)$$

$$= \left(\frac{\partial A_\mu}{\partial x_\nu} - \frac{\partial A_\nu}{\partial x_\mu}\right)(\Psi) \left(\text{since } \left[\frac{\partial}{\partial x_\mu}, \frac{\partial}{\partial x_\nu}\right] = 0\right)$$

Notice that this is, in fact, just the formula for the electromagnetic field, if the A_μ are interpreted as the potentials.

Now, suppose g is a gauge-transformation acting on the vector bundle E. Let ∇ be a linear connection for the vector bundle E, and let $D(g)(\nabla) = \nabla'$ be the transformed connection, given by (3.3). We want to compute the curvature tensor $\underset{\sim}{F}'$ of ∇' in terms of g and the curvature tensor $\underset{\sim}{F}$

QUANTUM FIELD THEORY 231

of ∇.

For $X, Y \in V(\ )$, $\Psi \in \Gamma(E)$,

$$\nabla_X'(\nabla_Y'\Psi) = D(g)(\nabla_{D(g^{-1})X} D(g^{-1})(\nabla_Y'(\Psi))$$

$$= D(g)(\nabla_{D(g^{-1})X} D(g^{-1})(D(g)\nabla_{D(g^{-1})Y} D(g^{-1})\Psi))$$

$$= D(g)\nabla_{D(g^{-1})X}\nabla_{D(g^{-1})Y} D(g^{-1})\Psi.$$

Hence,

$$\underset{\sim}{F}'(X,Y)(\Psi) = D(g)(\underset{\sim}{F}(D(g^{-1})X, D(g^{-1})Y))D(g^{-1})(\Psi) \quad (4.2)$$

Notice again that this is a typical "tensorial" transformation law. In fact, $\underset{\sim}{F}$ can be considered as a cross-section of a vector bundle E' over M. The fibre E'_p over a point $x \in M$ consists of the space of all skew-symmetric maps of $M_p \times M_p \to$ (space of linear transformations of E_p).

5. GAUGE INVARIANT COUPLINGS BETWEEN CROSS-SECTIONS AND CONNECTIONS

The tensorial nature of the curvature of a linear connection enables us to set up gauge-invariant differential equations, similar to the differential equations connecting the Dirac and electromagnetic fields.

Let E be a vector bundle over a manifold M, with a given group G of Gauge transformations acting on E. Let $\underset{\sim}{C}(E)$ be the space of linear connections for E. G acts on $\Gamma(E)$ and $\underset{\sim}{C}(E)$, as we have indicated above.

If E and E" are vector bundles over the same manifold M, let $\underset{\sim}{D}(E, E")$ be the space of differential operators: $\Gamma(E) \to \Gamma(E")$, i.e. an element Δ of $\underset{\sim}{D}(E, E')$ is a linear (but not necessarily an F(M)-linear) map $\Gamma(E) \to \Gamma(E')$ which, when expressed in terms of local coordinates of M and local cross-sections of E and E", is a differential operator in the ordinary sense. Such a Δ is a zero-th order differential operator, i.e. a tensor-field, if it is also F(M)-linear. If G acts as a gauge group on E and E', it also acts on $\underset{\sim}{D}(E, E")$:

$$D(g)(\Delta) = D(g)\Delta D(g^{-1}) \qquad (5.1)$$

for $\Delta \in \underset{\sim}{D}(E, E")$, $g \in G$.

Now, let E' be the vector bundle such that

QUANTUM FIELD THEORY 233

the curvature tensor of a connection $\nabla \in \underset{\sim}{C}(E)$ is
a cross-section of E', i.e. the operation of
applying the curvature operator defines a mapping:

$$\underset{\sim}{C}(E) \to \Gamma(E')$$

that intertwines the action of G.

To define the gauge-invariant couplings we
have in mind, we need the following data:

    a)   A mapping α: $\underset{\sim}{C}(E) \to \underset{\sim}{D}(E, E)$

    b)   A vector bundle E" over M on which G
also acts as a gauge-group.

    c)   A (non-linear) mapping

$$\alpha: \; \Gamma(E) \to \Gamma(E'')$$

    d)   A linear differential operator

$$\Delta \in \underset{\sim}{D}(E', E'').$$

We can now set up the equations we have in mind:

$$\gamma(\nabla)(\Psi) = 0$$

$$\underset{\sim}{D}(F_\nabla) = \alpha(\Psi) \tag{5.2}$$

Here, $\underset{\sim}{F}_\nabla$ is the curvature tensor of the connection ∇. Clearly, if the data a) - d) intertwines the action of G, and if

$$(\Psi, \nabla) \in \Gamma(E) \times \underset{\sim}{C}(E)$$

is a solution of (5.2), then so is

$$(D(g)\Psi, D(g)(\nabla)), \quad \text{for all} \quad g \in G.$$

This is what is meant by saying that the equations (5.2) are "gauge-invariant."

For example, in the Dirac-electromagnetic field case, these objects take the following form:

M is real 4-space, with coordinates x_μ ($0 \leqq \mu \leqq 3$). E is the product bundle M × V, where V is vector-space of Dirac spinors, i.e. a four dimensional vector space on which is given the (1/2, 0) $\oplus$ (0, 1/2)-representation of the Lorentz group. ∇ is determined by real-valued functions $A_\mu(x)$:

$$\nabla_{\frac{\partial}{\partial x_\mu}}(\Psi) = \frac{\partial \Psi}{\partial x_\mu} - ieA_\mu \Psi,$$

QUANTUM FIELD THEORY

where Ψ, a cross-section of E, is a map $M \to V$, e a real number, the charge of the electron. γ^μ are the Dirac-matrices acting on V as linear operators.

$$\gamma(\nabla)(\Psi) = \gamma^\mu \left(\frac{\partial}{\partial x_\mu} - iA_\mu \right)(\Psi) - im\Psi$$

(m is a real number).

α is the mapping: $\Psi(x) \to (e\bar{\Psi}(x)\gamma^\mu \Psi(x))$.

Δ is the differential operator

$$\underset{\sim}{F}_{\mu\nu} \to \frac{\partial \underset{\sim}{F}_{\mu\nu}}{\partial x_\nu} .$$

(Our notations are slight modifications of those given in [2], p. 84.)

(5.2) then takes the form:

$$\gamma^\mu \left(\frac{\partial}{\partial x_\mu} - ieA_\mu \right) \Psi = im\Psi$$

$$\underset{\sim}{F}_{\mu\nu} = \frac{\partial A_\mu}{\partial x_\nu} - \frac{\partial A_\nu}{\partial x_\mu}$$

$$\frac{\partial \underset{\sim}{F}_{\mu\nu}}{\partial x_\nu} = e\bar{\Psi}(x)\gamma^\mu \Psi$$

Notice that E" is also a product bundle, with its fibre a tensor product of V with R^4.

BIBLIOGRAPHY

1. R. Abraham, Foundations of Mechanics, W. A. Benjamin, New York, 1967.
2. J. D. Bjorken and S. Drell, Relativistic Quantum Fields, McGraw-Hill, New York, 1965.
3. J. M. Cook, Trans. Amer. Soc. $\underline{74}$, 222 (1953).
4. S. Helgason, Differential Geometry and Symmetric Spaces, Academic Press, New York, 1962.
5. R. Hermann, Lie Groups for Physicists, W. A. Benjamin, New York, 1966.
6. R. Hermann, Analytic Continuation of Group Representations, V. Comm. Math. Phys. $\underline{5}$, 157-190 (1967).
7. R. Hermann, Differential Geometry and the Calculus of Variation, Academic Press, New York, 1968.
8. H. Jost, The General Theory of Quantized Fields, American Math. Society, Providence, R.I., 1965.
9. D. Kastler, Introduction a l'electrodynamique, Dunod, Paris, 1961.
10. A. Kurosh, The Theory of Groups, Chelsea Pub. Co., New York, 1955.

11. L. Loomis and S. Sternberg, Advanced Calculus, Addison-Wesley, Reading, Mass., 1966.
12. R. L. Mills and C. N. Yang, Conservation of Isotopic Spin and Isotopic Gauge Invariance, Phys. Rev. $\underline{96}$, 191-5 (1954).
13. I. Segal, Mathematical Problems of Relativistic Physics, American Math. Society, Providence, R.I., 1963.
14. R. F. Streater and A. S. Wightman, PCT, Spin and Statistics, W. A. Benjamin, New York, 1964.

CHAPTER VIII

GENERALIZED FUNCTIONS ON MANIFOLDS

1. INTRODUCTION

The theory of generalized functions is of obvious importance both in quantum physics and in group representation theory. In this chapter we will discuss certain features of the theory that seem most relevant for these applications, but that are only covered in part by the standard treatises. As in Chapter I, our emphasis will be algebraic, following ideas of Dirac, rather than the traditional treatments in the mathematical literature emphasizing topological vector space theory. In addition, we will emphasize the connection with differential and integral geometry on manifolds, using [6] as foundation, and adopting the notations used there.

2. DIRAC SPACES

Let H be a complex vector space, with a positive-definite, Hermitian inner product:

$$(\Psi, \Psi') \to \langle\Psi|\Psi'\rangle \quad \text{for} \quad \Psi, \Psi' \in H. \quad (2.1)$$

Thus, in addition to bilinearity, the inner product (2.1) satisfies:

$$\langle c\Psi|\Psi'\rangle = c^*\langle\Psi|\Psi'\rangle$$

for $c \in C$ (= complex numbers)

c^* = complex conjugate of c.

$\langle\Psi|\Psi\rangle > 0$, unless $\Psi = 0$

$\langle\Psi|\Psi'\rangle = \langle\Psi'|\Psi\rangle^*$.

H, with such an inner product (not necessarily complete) is usually called a "pre-Hilbert space" in the mathematical literature. Following the convention in physics, we will call it simply a Hilbert space.

The *Dirac space* associated with H is the set of *all* linear functionals α: H → C, which we denote by $\underset{\sim}{D}$, or by $\underset{\sim}{D}_H$ if more than one such space is to be considered. $\underset{\sim}{D}$ is made into a complex vector space as follows:

a) $(\alpha_1 + \alpha_2)(\Psi) = \alpha_1(\Psi) + \alpha_2(\Psi)$

b) $(c\alpha_1)(\Psi) = c^* \alpha_1(\Psi)$

for $\alpha_1, \alpha_2 \in \underset{\sim}{D}$, $\Psi \in H$. (2.2)

For $\Psi \in H$, define $\alpha_\Psi \in \underset{\sim}{D}$ as follows:

$$\alpha_\Psi(\Psi') = <\Psi|\Psi'> \quad \text{for} \quad \Psi' \in H. \quad (2.3)$$

Notice that (2.2b) distinguishes $\underset{\sim}{D}$ from the usual definition of the "dual space" of H. Of course, (2.2b) is chosen to assure that the mapping: $\alpha \to \alpha_\Psi$ is a complex linear mapping: H → $\underset{\sim}{D}$. Then, we can use (2.3) to identify H with a subspace of $\underset{\sim}{D}$. This identification can be emphasized by adopting the following notations:

$$\langle\alpha|\Psi\rangle = \alpha(\Psi)$$

$$\langle\Psi|\alpha\rangle = \alpha(\Psi)^*$$

for $\Psi \in H, \alpha \in \underset{\sim}{D}$. $\hspace{2cm}$ (2.4)

A key problem is then the extension of various algebraic notions from H to $\underset{\sim}{D}$. For example, consider linear transformations: Given a linear transformation A: $H \to H'$ between two Hilbert spaces H and H', with associated Dirac spaces $\underset{\sim}{D}$ and $\underset{\sim}{D}'$, how can A be extended to a linear transformation: $\underset{\sim}{D} \to \underset{\sim}{D}'$? Of course, the problem is to do this extension in a "functorial" way -- meaning, roughly, that all algebraic relations are preserved by the extension process. One method of constructing this extension is to assume that an "adjoint" operator A^*: $H' \to H$ exists, such that:

$$\langle A\Psi|\Psi'\rangle = \langle\Psi|A^*\Psi'\rangle \quad \text{for} \quad \Psi, \Psi' \in H. \quad (2.5)$$

The extension of A to a map $\underset{\sim}{D} \to \underset{\sim}{D}'$, also denoted by A, is then defined as follows:

$$\langle A\alpha|\Psi'\rangle = \langle\alpha|A^*\Psi'\rangle \quad \text{for} \quad \alpha \in \underset{\sim}{D}, \Psi' \in H.$$
$$(2.6)$$

GENERALIZED FUNCTIONS ON MANIFOLDS

The "functorial" nature of this extension process is reflected at an explicit level by the following result (proof left to the reader).

LEMMA 2.1. Suppose $A_1: H_1 \to H_2$, $A_2: H_2 \to H_3$ are linear transformations, such that A_1^*, A_2^* exist. Set: $A = A_2 A_1$. Then, $A^* = A_1^* A_2^*$. With this choice of adjoint operator, and the corresponding extensions of A to a map: $\underset{\sim}{D}_1 \to \underset{\sim}{D}_3$ via (2.6), we have:

$$A(\alpha) = A_2(A_1(\alpha)) \quad \text{for} \quad \alpha \in \underset{\sim}{D}. \qquad (2.7)$$

The formalities are relevant in group representation theory. Suppose that G is a Lie group, and that $g \to D(g)$ is a representation of G by linear operators on H. Assume that $D(g)^*$ exists, for each $g \in G$. We can then perform the following operations:

    a) Extend $D(G)$ to be operators on $\underset{\sim}{D}$. (2.7) then asserts that this is also a representation of G.

    b) Define a new representation $D'(g)$ as follows:

$$D'(g) = D(g^{-1})^*. \qquad (2.8)$$

Then, $D'(g)^*$ can be taken to be $D(g^{-1})$, hence D' also can be extended to give a representation of G by operators on $\underset{\sim}{D}$.

Let $\underset{\sim}{G}$ be the Lie algebra of G. Suppose that H can be made into a topological vector space in such a way that the operators $D(X)$, $D'(X)$, for $X \in \underset{\sim}{G}$, exist:

$$D(X)(\Psi) = \lim_{t \to 0} \frac{D(\exp(tX))(\Psi) - \Psi}{t}$$

$$D'(X)(\Psi) = \lim_{t \to 0} \frac{D'(\exp(tX))(\Psi) - \Psi}{t} \qquad (2.8)$$

Let us examine the conditions that $D'(X) = D(-X)^*$: First, let us assume that the convergence used on the right hand side of (2.8) implies weak convergence in the Hilbert space sense, i.e.

If $\Psi_1, \Psi_2, \ldots$ is a sequence of elements of H that converges to an element Ψ of H in the given topology, then

$$\lim_{n \to \infty} \langle \Psi_n | \Psi' \rangle = \langle \Psi | \Psi' \rangle$$

for all $\Psi' \in H$. $\qquad (2.9)$

Then, using (2.9),

$$\langle\Psi|D'(X)\Psi\rangle = \lim_{t \to 0} \frac{1}{t} \langle\Psi|D'(\exp(tX))\Psi' - \Psi\rangle$$

$$= \lim_{t \to 0} \frac{1}{t} (\langle D(\exp(-tX))\Psi|\Psi'\rangle - \langle\Psi|\Psi'\rangle)$$

$$= \langle D(-X)\Psi|\Psi'\rangle, \text{ i.e. } D(X)^* = -D'(X) \tag{2.10}$$

If further the action of $D(G)$ on the topological vector vector space H is a topological-group action, then $D(G)$ and $D'(G)$ are representations of the Lie algebra $\underset{\sim}{G}$, and (2.10) assures that both of these representations extend to representations on $\underset{\sim}{D}$.

Of course, in adopting the formalism of incomplete Hilbert spaces we have abandoned the crutches provided by the standard theorems of Hilbert space theory and functional analysis, e.g. the spectral theorem and the theory of self-adjoint operators and the self-adjoint extensions of symmetric operators, the Hahn-Banach theorem, etc.

However, we hope to replace these techniques (which are anyway usually too general to capture in a simple way the most interesting features of most of the special situations) with more algebraic rules that are better suited to the circle of problems encountered in group representation theory and quantum physics.

3. DIRAC SPACES MANIFOLDS AND THEIR BEHAVIOR UNDER MAPPINGS

Let M be a manifold. (We refer to [6] for the background in manifold theory used here. All data is of differentiability class C^∞, unless mentioned otherwise.) Let H be a Hilbert space, in the sense described in Section 2. Assume that, as a vector space, H is a subspace of the space F(M) of C^∞, complex-valued functions on M.

Let us assume that the inner product on H is given by integration over M. To consider the simplest situation, suppose that M is orientable, with dp a volume-element differential form on M. (That is, dp is a differential form of degree equal to the dimension of M which is non-zero at each point of M.) Denote a typical point of M by p. Then,

GENERALIZED FUNCTIONS ON MANIFOLDS 247

$$\langle \Psi_1 | \Psi_2 \rangle = \int_M \Psi_1(p)^* \Psi_2(p) dp$$

for $\Psi_1, \Psi_2 \in H$. (3.1)

Associating H with a space of functions enables us to define the "support" of elements of $\underset{\sim}{D}$, as is customary in the theory of distributions. Explicitly: Consider an element $\alpha \in D$. α is said to *vanish in an open set of M* if

$$\langle \alpha | \Psi \rangle = 0$$

for each $\Psi \in H$ which vanishes *outside* of this open set. The complement in M of the union of all the open subsets of M in which α vanishes is a closed subset of M, called the *support* of α, and denoted by:

$$\text{supp}(\alpha).$$

If N is a subset of M, α is said to be *equal to an element* Ψ *of H outside of N* if:

$$\text{supp}(\alpha - \alpha_\Psi) \subset N.$$

Suppose that H' is a similarly-defined Hilbert space of functions on a manifold M'. Consider a map $\phi: M \to M'$. ϕ defines a dual map

$$\phi^*: F(M') \to F(M):$$

$$\phi^*(\Psi')(p) = \Psi'(\phi(p)) \qquad (3.2)$$

for $\Psi' \in F(M')$.

(Of course, there is some possibility of confusion between this use of the * notation and that used in Section 2 for Hermitian adjoint. Since both notations are so standard in the differential-geometric and physics literature, we will just have to leave it up to the reader to be careful of possible confusion.) Let us then define $A: H' \to H$ as the restriction of ϕ^* to H', i.e.

$$A(\Psi') = \phi^*(\Psi') \quad \text{for} \quad \Psi' \in H'. \qquad (3.3)$$

The main problem can now be stated as follows: How can A be extended to a mapping: $D' \to D$ of Dirac spaces? We shall now indicate how the methods of integral geometry can be used to answer this problem, at least under certain hypotheses.

Let us suppose that:

$$\dim M \geq \dim M'; \quad \phi(M) = M'$$

For $p \in M$, let M_p denote the tangent space to M at p. Let $\phi_*: M_p \to M'_{\phi(p)}$ denote the linear map on tangent vectors induced by ϕ. p is said to be a *nonsingular point* for ϕ if:

$$\phi_*(M_p) = M'_{\phi(p)}.$$

A point $p' \in M'$ is a *regular value for* ϕ if $\phi^{-1}(p')$ consists of non-singular points. By Sard's theorem [10], the set of non-regular values in M' has measure zero.

Let dp' be a volume-element differential form on M'. Denote by $\phi^*(dp')$ its pull-back relative to ϕ. Let θ be a differential form on M such that:

$$\text{degree } \theta + \dim M' = \dim M.$$

Thus, $\theta \wedge \phi^*(dp')$ is a differential form of maximal degree on M. Let us suppose that M is an oriented manifold, so that forms of this maximal degree can be integrated over M. Then, one of the two basic integral geometric formulae is:

$$\int_M f(p)(\theta \wedge \phi^*(dp')) = \int_{M'} \left(\int_{\phi^{-1}(p')} f(p)\theta \right) dp' \quad (3.4)$$

The convention inherent in the right hand side of (3.4) can be explained as follows: If p' is a regular value of M', the fibre $\phi^{-1}(p')$ is a submanifold of M of dimension equal to the degree of the differential form fθ. (f is a function on M, i.e. an element of F(M)).

$$\int_{\phi^{-1}(p')} f(p)\theta$$

denotes the integral of this form over the fibre.

$$p' \to \int_{\phi^{-1}(p')} f(p)\theta$$

is then a function defined on the set of regular values in M'. Its integral over M', i.e. the right hand side of (3.4), is now well-defined by standard measure theory, since the complement in M' of the set of regular values is of measure zero. The geometric background of this formula is explained

GENERALIZED FUNCTIONS ON MANIFOLDS 251

in [6]. Its proof in complete generality was given (in a different language) by Federer [2].

Let us now return to computing the adjoint operator A^* of the operator A defined by (3.3). Suppose that the form θ is chosen so that:

$$dp = \theta \wedge \phi^*(dp') \qquad (3.5)$$

For $\Psi \in H$, $\Psi' \in H'$,

$$<A^*(\Psi)|\Psi'> = <\Psi|A(\Psi')> = \int_M \Psi(p)^* \Psi'(\phi(p)) dp$$

$$= \text{, using (3.4-5),}$$

$$\int_{M'} \left(\int_{\phi^{-1}(p')} \Psi(p)^* \Psi'(\phi(p)) \theta \right) dp'$$

$$= \int_{M'} \left(\int_{\phi^{-1}(p)} \Psi(p) \theta \right)^* \Psi'(p') dp'.$$

We see from this that:

$$A^*(\Psi)(p') = \int_{\phi^{-1}(p')} \Psi(p) \theta. \qquad (3.6)$$

i.e. A^* is "integration over the fibres."

Of course, whether (3.6) actually defines an

element of H' depends on how H' is precisely defined, and on the more subtle properties of ϕ, e.g. whether the right-hand side of (3.6), defined, strictly speaking only for the regular p' ϵ M', can be extended to function over all of M -- the close connection with the theory of singularities of mappings is the most interesting part of the whole problem.

Suppose that A^* satisfying (3.3) can be defined. Then, using (2.6), A can be extended to give a map $\underset{\sim}{D} \rightarrow \underset{\sim}{D}'$. In particular, if $\delta_{p_0'}$ is the Dirac delta function associated with a point p_0' of M' (i.e. $<\delta_{p_0'}|\Psi'> = \Psi(p_0')$), then $A(\delta_{p_0'})$ is a element of $\underset{\sim}{D}$. If p_0' is a regular element of M', (3.6) holds, and $A(\delta_{p_0'})$ can be computed as follows:

$$<A(\delta_{p_0'})|\Psi> \quad \int_{\phi^{-1}(p_0')} \Psi(p)\theta, \qquad (3.7)$$

i.e. $A(\delta_{p_0'})$ is the "delta-function spread out over the submanifold $\phi^{-1}(p_0')$," which has been extensively treated by Gel'fand and Shilov [3]. We notice then that this general "Dirac-space" point of view, together with the general insights of integral geometry, gives a unified formalism for many of the special calculations treated by Gel'fand and Shilov, which in turn is of great importance for numerous problems of partial differential equations and group representation theory.

The next order of business is to understand, using differential-geometric ideas, how the presence of singularities in the mapping ϕ affects the situation.

4. GENERALIZED FUNCTIONS ASSOCIATED WITH MAPPINGS OF PSEUDORIEMANNIAN MANIFOLDS

Let M and M' be manifolds, let $\phi: M \to M'$ be a mapping from M to M'. Our aim in this section is to consider the case where ϕ may have singularities. However, we will not proceed with the methods of the general theory of singularities of mappings, but will consider certain differential-geometric hypotheses concerning the situation. (These sort of hypotheses seem to be relevant to many of the examples that are of interest in the theory of generalized functions.)

First, we will assume that both M and M' have

fixed pseudoriemannian metrics. Recall that this means that to each point $p \in M$, $p' \in M'$, there is assigned a symmetric, bilinear form, to be denoted by $\langle\ ,\ \rangle$, on the tangent spaces M_p and $M'_{p'}$. We will assume that these forms are definite, but not necessarily positive-definite. We will also suppose that M and M' are orientable manifolds, which, together with the metrics, enables us to define volume-element differential forms dp and dp' on both M and M'. For example, we will recall the definition of dp.

Suppose dim $M = m$. A set $(\omega_1, \ldots, \omega_m)$ of 1-differential forms is said to be an *orthonormal basis* for the metric if:

$$\langle v, v \rangle = \pm \omega_1(v)^2 \pm \ldots \pm \omega_m(v) \qquad (4.1)$$

for each tangent vector v to M.

(Of course, the number of plus and minus signs in (4.1) is fixed by the index of the symmetric bilinear form $\langle\ ,\ \rangle$.) Given such an orthonormal basis, one can consider the m-form:

$$\omega_1 \wedge \ldots \wedge \omega_m. \qquad (4.2)$$

Since any two such orthonormal bases differ by an orthogonal matrix, which has determinant ±1, this m-form is, up to sign, independent of the orthonormal basis.

An *orientation* on M is, by definition [6], determined by choice of an everywhere non-zero m-form on M. Suppose such an orientation is given. Then, let us say that the orthonormal basis is *positively oriented* if the m-form (4.2) is a positive multiple of this given orientation-form. We can then define the m-form dp globally over M by requiring that:

$$dp = \omega_1 \wedge \cdots \wedge \omega_m \qquad (4.3)$$

for each positively oriented orthonormal basis $(\omega_1, \ldots, \omega_m)$.

Another basic feature of pseudoriemannian metrics is the induced metric form on submanifolds. Suppose that N is a submanifold of M. As explained in [6], we will, for the sake of simplicity of notation, identify N with a subset of M, and identify the tangent space N_p to each point p ε M with a subspace of M_p. Then, the form < , >

induces by restriction to N_p a symmetric bilinear form on N_p. Of course, this form need not be definite (except if the form $< \, , \, >$ is positive, i.e. the metric is Riemannian). Let us say that N is a *non-characteristic* submanifold if the form $< \, , \, >$ restricted to each tangent space N_p is definite. Then, of course, the form $< \, , \, >$ defines a pseudo-riemannian metric on N, hence, if N is oriented, a volume element differential form on N, as explained above.

We will now make certain assumptions about the map $\phi: M \to M'$:

M has an open, dense subset, denoted by M_R, such that

$\phi(M_R)$ is a non-characteristic submanifold of M'. (4.4)

$\phi^{-1}(\phi(M_R)) \subset M_R$, i.e.

M_R is the union of fibres of ϕ. (4.5)

(The points of M_R will be called the *regular points* of the mapping ϕ. The points of $\phi(M_R)$ will be called the *regular image points* of ϕ.)

ϕ restricted to M_R, a map: $M_R \to \phi(M_R)$, is a maximal rank mapping to which the formula (3.4) applies. (4.6)

$\phi(M)$ is the closure of $\phi(M_R)$ in M'. (4.7)

For each $p' \in \phi(M)$, the fibre $\phi^{-1}(p')$ is an oriented submanifold of M. (4.8)

For each $p' \in \phi(M_R)$, $\phi^{-1}(p')$ is a non-characteristic submanifold of M. (4.9)

Suppose given now a vector space H of C^∞, complex-valued functions on M. Our main problem can be defined as follows: Can we define a vector space H' of functions on $\phi(M)$, and a mapping $A^*: H \to H'$ that is a generalization of the mapping denoted by A^* in Section 3. In fact, our hypotheses are strong enough to give a crude affirmative answer to this question.

For $p \in M$, denote by $N(p)$ the fibre $\phi^{-1}(\phi(p))$ of ϕ passing through p. By (4.8), it is a submanifold of M. (Of course, this is the weakest point in our assumptions -- it is very easy to construct simple mappings with very pathological fibres. One might hope that the "Hausdorff measure" techniques used by Federer [2], will be useful in weakening this hypothesis, particularly if they can be adopted to the "pseudoriemannian" geometric situation.)

Suppose now that $p \varepsilon M_R$, and that $\Psi \varepsilon H$. $A^*(\Psi)$ is to be a function on $\phi(M)$. Let dp' be the volume element form on $\phi(M_R)$ defined by the pseudo-riemannian metric on M'. Consider a form θ such that:

$$dp = \theta \wedge dp' \qquad (4.10)$$

Then, θ restricted to $N(p)$ will be a multiple $j(p)$ of the volume element-element form on $M(p)$ defined by the pseudoriemannian metric. θ is not uniquely determined by (4.10), but $j(p)$ is, hence we can consider $p \to j(p)$ as a real-valued function defined over M_R. ($j(p)$ reduces to the classical Jacobian function if: $\dim M_R = \dim \phi(M_R)$, which is the reason for choosing this notation for it.) This suggests that we define:

$$A^*(\Psi)(\phi(p)) = \int_{N(p)} \Psi(q) j(q) dq \qquad (4.10)$$

for $p \in M_R$.

In (4.10), "q" denotes a typical point of $N(p)$. "dq" denotes the volume element form defined by the induced pseudoriemannian metric on $N(p)$.

Notice now that (4.10) is in a form which is more suitable for the job of defining $A^*(\Psi)(p')$ for an arbitrary point $p' \in \phi(M)$. It decomposes the possible types of singularities of this function into what seem to be the most important "geometric" components, the singularities in the function $p \to j(p)$ due to the possible breakdown in the maximal-rank condition for ϕ for points p on the boundary of M_R, and the possible tendency of the metric on the fibres $N(p)$ to become "characteristic" as p moves toward the boundary of M_R.

Another general approach is suggested by (4.10). Let H' be a Hilbert space of functions $\Psi'(p')$ defined for $p' \in \phi(M)$, with the inner product defined by:

$$<\Psi_1'|\Psi_2'> = \int_{\phi(M_R)} \Psi_1'(p')^* \Psi_2'(p') dp' \quad (4.11)$$

Let us attempt to exhibit $A^*(\Psi)$ as an element of

the Dirac space $\underset{\sim}{D}'$ associated with H'. Formally, we would have:

$$<A^*(\Psi)|\Psi'> = \int_{\phi(M_R)} (\int_{N(\phi^{-1}(p'))} \Psi(q)^*j(q))\Psi'(p')dp'$$

Now, the integral (4.12) might not exist for all $\Psi' \in H'$. However, it will certainly exist if Ψ and Ψ' are of compact support, and if, for example, the support of Ψ' lies in $\phi(M_R)$. Thus, one might attempt to extend the bilinear function on these function spaces, defined by the right hand side of (4.12), to wider classes of functions. This is, in fact, the basic problem in the theory of "regularization of divergent integrals." Now, one can perceive, in the examples worked out by Gel'fand and Shilov [3], a beginning towards a general theory of regularization. (Of course, many problems of renormalization in quantum field theory are of this nature also [1, 4].) What is clearly needed is systematic work in directions that will ultimately unify the geometric-topological approach to the problem with the regularization approach. For example, one would like a theorem which

GENERALIZED FUNCTIONS ON MANIFOLDS

asserted that a certain type of regularization is related in some way to certain types of singularities of mappings.

As a very small step towards this program, we will now present a development of some of the topics treated by Gel'fand and Shilov [3], in terms of manifold theory.

5. GENERALIZED FUNCTIONS DEFINED BY HYPERSURFACES

Let M be an oriented manifold, and let N be an oriented hypersurface, i.e. an oriented submanifold of M of codimension one. Let f be a real-valued C^∞ function on M such that:

$$f(N) = 0 \qquad (5.1)$$

$$df \neq 0 \quad \text{at every point of N.} \qquad (5.2)$$

Let dp be a volume-element differential form for M. An (n-1)-form θ_f defined in a neighborhood of N is said to be *dual* to f if:

$$dp = \theta_f \wedge df \qquad (5.3)$$

Then, if Ψ is a complex-valued function defined in a neighborhood of N,

$$\int_N \Psi \theta_f \qquad (5.4)$$

is, as is readily verified, independent of the choice of θ_f satisfying (5.3). (5.4) defines a linear functional on functions, that we consider as a "generalized function," and denote by $\delta(f)$, i.e.

$$<\delta(f)|\Psi> = \int_N \Psi \theta_f \qquad (5.5)$$

Let X be a vector field on M. Then, $\Psi \to X(\Psi)$ is a derivation on the ring of functions on M. Let (div X) be the function on M such that:

$$X(dp) = (\text{div } X)dp \qquad (5.6)$$

(The left hand side of (5.6) denotes the Lie derivative of the form dp by the vector field X. Note that we are following the differential-geometric notations described in [6].) The adjoint map X^*, considered as a map from functions to functions, can be defined so as to satisfy:

GENERALIZED FUNCTIONS ON MANIFOLDS

$$<X\Psi_1|\Psi_2'> = <\Psi_1|X^*\Psi_2> \qquad (5.7)$$

for two functions Ψ_1, Ψ_2 which vanish at infinity on M, where the inner product is defined as follows:

$$<\Psi_1|\Psi_2> = \int_M \Psi_1(p)^* \Psi_2(p) dp. \qquad (5.8)$$

Explicitly then,

$$<X\Psi_1|\Psi_2> = \int_M X(\Psi_1^*)\Psi_2 dp$$

$$= - \int_M \Psi_1^*(X(\Psi_2) + \Psi_2 \, \text{div } X) dp$$

Comparing this with (5.7) gives:

$$X^*(\Psi) = - X\Psi - (\text{div } X)\Psi \qquad (5.9)$$

Let H be a Hilbert space of C^∞ functions on M, with inner product defined by (5.8). Let $\underset{\sim}{D}$ be its Dirac space. According to the general principle of Section 2, we can extend X to be a mapping: $\underset{\sim}{D} \to \underset{\sim}{D}$ using the rule:

$$\langle X(\alpha)|\Psi\rangle = \langle\alpha|X^*\Psi\rangle$$
$$= -\langle\alpha|X\Psi + (\operatorname{div} X)\Psi\rangle \qquad (5.10)$$

In particular, we may use this rule to apply X to the element $\delta(f) \in \underset{\sim}{D}$ defined by (5.5). We now have:

<u>LEMMA 5.1</u>. If $X(f) = 0$, then $X(\delta(f)) = 0$.

<u>Proof</u>. For $\Psi \in H$,

$$\langle X(\delta(f))|\Psi\rangle = -\langle\delta(f)\ X(\Psi) + \operatorname{div} X\Psi$$
$$= -\int_N (X(\Psi) + \Psi \operatorname{div} X)\theta_f$$

Using (5.3), we have:

$$(\operatorname{div} X)dp = X(dp) = X(\theta_f) \wedge df,$$

or $\quad (\operatorname{div} X\theta_f - X(\theta_f)) \wedge df = 0$.

Using (5.1), we see that:

$$\int_N (\operatorname{div} X)\theta_f = \int_N X(\theta_f)$$

Hence,

$$\langle X(\delta(f))|\Psi\rangle = -\int_N X(\Psi)\theta_f + \Psi X(\theta_f)$$

$$= -\int_N X(\Psi\theta_f)$$

Now,

$$X(\Psi\theta_f) = X \lrcorner\, d(\Psi\theta_f) + d(X \lrcorner\, \Psi\theta_f).$$

Since $X(f) = 0$, X is tangent to N, hence the form $X \lrcorner\, d(\Psi\theta_f)$ is zero when restricted to N. By Stokes' theorem and our assumption that Ψ vanishes at "infinity" (i.e. at the boundary) of N, the integral

$$\int_N d(X \lrcorner\, \Psi\theta_f)$$

vanishes also. This proves the lemma.

Lemma (5.1) enables us to define the "derivatives" $\delta^1(f)$, $\delta^2(f)$,... of the "delta function $\delta(f)$. To do this, choose a vector field Y such that:

$$Y(f) = 1. \tag{5.11}$$

Set:

$$\delta^1(f) = Y(\delta(f)); \quad \delta^2(f) = Y(\delta^1(f)) \qquad (5.12)$$
etc.

LEMMA 5.2. $\delta^1(f)$, $\delta^2(f)$, etc., as given by (5.12), are independent of the Y vector field Y used to define it, i.e. if Z is a vector field such that: $Z(f) = 0$, then $Z(\delta^{(k)}(f)) = 0$ for all k.

Proof. If Y' is another vector field satisfying (5.11), we have that: $Z = Y-Y'$ satisfies $Z(f) = 0$.

By Lemma (5.1), $0 = Z(\delta(f)) = Y(\delta(f)) - Y'(\delta(g))$. Let us prove that: $Z(\delta^1(f)) = 0$. The proof the rest will be similar:

$$Z(\delta^1(f)) = ZY(\delta(f))$$
$$= [Z, Y](\delta(f)) + YZ(\delta(f)).$$

Now, $[Z, Y](f) = ZY(f) - YZ(f) = 0$. Lemma (5.1) now implies that $Z(\delta^1(f))$ is indeed zero.

LEMMA 5.3. If X is a vector field on M, then the "chain rule" holds in the following form:

GENERALIZED FUNCTIONS ON MANIFOLDS

$$X(\delta^k(f)) = X(f)\delta^{(k+1)}(f). \qquad (5.13)$$

Proof. Set:

$$Z' = X - X(f)Y.$$

Then, $Z(f) = 0$, hence, by Lemma (5.2),

$$Z(\delta^k(f)) = 0.$$

θ_f can also be calculated in terms of the Y satisfying (5.11). In fact, using (5.3):

$$Y \lrcorner dp = (Y \lrcorner \theta_f) \wedge df$$
$$+ (-1)^{m-1}\theta_f \wedge (Y \lrcorner df).$$

Since $Y \lrcorner df = Y(f) = 1$, and $df = 0$ on N, we have that:

$$Y \lrcorner dp - (-1)^{m-1}\theta_f \quad \text{restricted to N.} \quad (5.14)$$

$$(m = \dim M)$$

Suppose now that M has a pseudoriemannian metric, $< \ , \ >$. If $p \to f(p)$ is a real-valued function on M, let grad f be the vector field such that:

$$\langle \text{grad } f, X \rangle = X(f)$$

for each vector field X on M.

If dp is the volume element on M associated with the matric, then, by definition, div (grad f) is the Laplacian $\Delta(f)$ of f [6]. Set:

$$Y = \frac{\text{grad } f}{\langle \text{grad } f, \text{grad } f \rangle} \qquad (5.16)$$

Note that: $Y(f) = 1$. Also, note that:

Y is perpendicular to the submanifold N

(5.17)

Proof. Suppose that X is a vector field that is tangent to N, i.e.

$$X(f) = 0 \quad \text{on} \quad N.$$

Then,

$$\langle Y, X \rangle = \frac{\langle \text{grad } f, X \rangle}{\langle \text{grad } f, \text{grad } f \rangle}$$

$$= X(f)/\langle \text{grad } f, \text{grad } f \rangle = 0.$$

This proves (5.17).

$$\langle Y, Y \rangle = \frac{1}{\langle \text{grad } f, \text{grad } f \rangle} \qquad (5.18)$$

Thus, $Y' = Y \langle \text{grad } f, \text{grad } f \rangle^{1/2}$ is a vector field of unit length that is perpendicular to N. Hence,

$$Y' \lrcorner \, dp$$

is a volume-element differential form that is, by definition, the volume element defined by the induced pseudoriemannian metric on M. Let us denote this volume element by: dq. Thus, by (5.14),

$$\langle \delta(f) | \Psi \rangle = (-1)^{m-1} \int_N \langle \text{grad } f, \text{grad } f \rangle^{-\frac{1}{2}} \Psi(q) dq.$$
$$(5.19)$$

In particular, suppose that f is a solution of the Hamilton-Jacobi equation associated with the metric [7]. This can be characterized in two equivalent ways: either by requiring that the integral curves of grad f be geodesics, or that there is a relation of the form:

$$\langle \text{grad } f, \text{grad } f \rangle = F(f). \qquad (5.20)$$

If f satisfies (5.20), note that (5.19) takes the form:

$$\langle \delta(f) | \Psi \rangle = (-1)^{m-1} F(0)^{\frac{1}{2}} \int_N \Psi(q) dq \qquad (5.21)$$

LEMMA 5.4. The following identities hold:

a) $f\delta(f) = 0$
b) $f\delta^1(f) + \delta(f) = 0$
$$\vdots$$
$f\delta^k(f) + k\delta^{(k-1)}(f) = 0 \qquad (5.22)$

Proof. For $\Psi \in H$,

$$\langle f\delta(f) | \Psi \rangle = \langle \delta(f) | f\Psi \rangle = 0,$$

which proves a).

To prove b), choose a vector field Y such that

$$Y(f) = 1.$$

Then, $\delta^1(f) = Y(\delta(f))$. Hence,

$$f\delta^{(1)}(f) = fY(\delta(f)) = Y(f\delta(f)) - Y(f)\delta(f)$$
$$= -\delta(f).$$

This proves b).

The rest of the identities are proved similarly. (Note that they are given on p. 233 of [3].)

In summary, we have presented in this section a treatment using manifold theory of some of the fundamental facts concerning the generalized functions defined by hypersurfaces.

6. GENERALIZED FUNCTIONS GENERATED BY FLOWS

Keep the notations of Section 5. Suppose now that t is a real parameter, with N^t, f^t dependent on t. Our problem: How does one compute

$$\frac{d}{dt} \delta(f^t) \qquad (6.1)$$

For this purpose, it is desirable to adopt a slightly more abstract viewpoint. Suppose that N is a manifold, of dimension (m-1), and $t \to \alpha_t$ is a one-parameter family of submanifold maps: $N \to M$. (Thus, for intuitive purposes, N^t may be identified with $\alpha_t(N)$.) Suppose that f^t is a one-parameter family of real-valued functions on M such that:

$$\alpha_t^*(f^t) = 0 \quad \text{for all t.}$$

Let Y^t be a one-parameter family of vector fields such that:

$$Y^t(f^t) = 1.$$

Thus, as we have seen in Section 5,

$$<\delta(f^t)|\Psi> = \int_N \alpha_t^*(\theta^t \Psi) \qquad (6.2)$$

where θ^t is the (m-1)-form such that

$$dp = \theta^t \wedge df^t$$

Of course, we can now proceed to calculate (6.1) using (6.2). However, we will not be interested for the moment in the most general case, but will suppose that the deformation α_t is generated by a *flow* on M, i.e. a one-parameter family $t \to \phi_t$

GENERALIZED FUNCTIONS ON MANIFOLDS

of diffeomorphisms of M such that:

$$\alpha_t(q) = \phi_t(\alpha_0(q)) \quad \text{for} \quad q \in N, \text{ all } t. \tag{6.3}$$

Let $t \to Z^t$ be the one parameter family of vector fields on M that is the *infinitesimal generator* of the flow. Z^t is characterized by the following condition:

$$\frac{\partial}{\partial t} \phi_t^*(\omega) = \phi_t^*(Z^t(\omega)) \tag{6.4}$$

for each differential form ω on M.

Combining (6.2) and (6.3) gives:

$$<\delta(f^t)|\Psi> = \int_N \alpha_0^* \phi_t^*(\theta^t \Psi)$$

Hence:

$$<\frac{\partial \delta(f^t)}{\partial t}|\Psi> = \int_N \alpha_0^* \phi_t^*(Z^t(\theta^t \Psi)) \tag{6.5}$$

Now, let us suppose that f^t is chosen so that:

$$Z^t(f^t) + \frac{\partial f^t}{\partial t} = 0 \tag{6.6}$$

Then,

$$\frac{\partial}{\partial t}(\alpha_t^*(f^t)) = \frac{\partial}{\partial t}(\alpha_0^* \phi_t^*(f^t))$$

$$= \alpha_0^* \phi_t^*(Z^t(f^t) + \frac{\partial f^t}{\partial t}) = 0.$$

The function $t \to f^t(\phi_t(\alpha_0(q)))$ is constant in t, for fixed q. If we assume then that: $f^0(\alpha_0(N)) = 0$, we will have: $f^t(\alpha_t(N)) = 0$ for all t, as required. Now, (6.5) shows that the following relation may be assumed between Z^t and Y^t.

$$Z^t = -(\frac{\partial f^t}{\partial t}) Y^t \qquad (6.7)$$

With relation (6.7), let us compute (6.5)

$$Z^t(\theta^t) = -\frac{\partial f^t}{\partial t} Y^t(\theta^t) - d(\frac{\partial f^t}{\partial t})^\wedge (Y^t \lrcorner \theta^t).$$

But, $\alpha_t^*(Y \lrcorner \theta^t) = 0$. Hence:

$$\langle \frac{\partial \delta(f^t)}{\partial t} | \Psi \rangle = \int_N \alpha_0^* \phi_t^* (-\frac{\partial f^t}{\partial t}(Y^t(\theta^t)$$
$$+ \theta^t Y^t(\Psi)))$$

As we have seen in Section 5, we may choose: θ^t

GENERALIZED FUNCTIONS ON MANIFOLDS

so that:

$$\theta^t = (-1)^{m-1} Y^t \lrcorner \, dp \qquad (6.8)$$

(m = dim M).

Then,

$$Y^t(\theta^t) = (-1)^{m-1} Y^t \lrcorner \, Y^t(dp)$$

$$= (\operatorname{div} Y^t)\theta^t \qquad (6.9)$$

Hence,

$$\langle \frac{\partial \delta(f^t)}{\partial t} | \Psi \rangle = -\int_N \alpha_0^* \phi_t^* (\frac{\partial f^t}{\partial t}((\operatorname{div} Y^t)\theta^t \Psi$$

$$+ \theta^t Y^t(\Psi))) \qquad (6.10)$$

Let us compare this with (5.12):

$$\langle \delta^1(f^t) | \Psi \rangle = - \langle \delta(f^t) | Y^t(\Psi) + \operatorname{div}(Y^t)\Psi \rangle$$

$$= - \int_N \alpha_0^* \phi_t^* (\theta^t(Y^t(\Psi) + (\operatorname{div} Y^t)\Psi)$$

Also,

$$\langle \frac{\partial f^t}{\partial t} \delta^1(f^t) | \Psi \rangle = \langle \delta^1(f^t) | \frac{\partial f^t}{\partial t} \Psi \rangle$$

$$= - \int_N \alpha_0^* \phi_t^* (\theta_t(\frac{\partial f^t}{\partial t} Y^t + \Psi)$$

$$+ Y^t(\frac{\partial f^t}{\partial t}) \Psi + \frac{\partial f^t}{\partial t} (\text{div } Y^t) \Psi)$$

Now,

$$Y^t(\frac{\partial f^t}{\partial t}) = \frac{\partial}{\partial t} (Y^t(f^t)) - \frac{\partial Y^t}{\partial t} (f^t)$$

$$= - \frac{\partial}{\partial t} (Z^t(f^t))/(\frac{\partial f^t}{\partial t})) - \frac{\partial Y^t}{\partial t} (f^t)$$

$$= 0 - (\frac{\partial Y^t}{\partial t})(f^t)$$

Hence, we have:

$$\frac{\partial \delta(f^t)}{\partial t} = \frac{\partial f^t}{\partial t} \delta^1(f^t) + \frac{\partial Y^t}{\partial t} (f^t) \delta(f^t)$$

Now, $Y^t(f^t) = 1$, hence, differentiating this relation with respect to t, gives:

$$\frac{\partial Y^t}{\partial t} (f^t) = - Y^t(\frac{\partial f^t}{\partial t}),$$

hence we have proved:

GENERALIZED FUNCTIONS ON MANIFOLDS

THEOREM 6.1.

$$\frac{\partial}{\partial t} \delta(f^t) = \delta^1(f^t) \frac{\partial f^t}{\partial t} - Y^t(\frac{\partial f^t}{\partial t})\delta(f^t) \quad (6.11)$$

We can compute $(\partial^2/\partial t^2)\delta(f^t)$ in a similar way:

$$\frac{\partial}{\partial t} \delta^1(f^t) = \frac{\partial}{\partial t} (Y^t(\delta(f^t)) = (\frac{\partial Y^t}{\partial t})(\delta(f^t))$$

$$+ Y^t(\delta^1(f^t) \frac{\partial f^t}{\partial t} - Y^t(\frac{\partial f^t}{\partial t})\delta(f^t))$$

$$= \delta^1(f^t)(\frac{\partial Y^t}{\partial t})(f^t) + \delta^2(f^t) \frac{\partial f^t}{\partial t}$$

$$+ \delta^1(f^t)(\frac{\partial f^t}{\partial t}) - Y^t(Y^t(\frac{\partial f^t}{\partial t}))\delta(f^t)$$

$$- Y^t(\frac{\partial f^t}{\partial t})\delta^1(f^t)$$

$$= \delta^2(f^t) \frac{\partial f^t}{\partial t} - Y^t(Y^t(\frac{\partial f^t}{\partial t}))\delta(f^t)$$

$$+ \frac{\partial Y^t}{\partial t}(f^t)\delta^1(f^t) \quad (6.12)$$

These formulas do not seem to be given by Gel'fand and Shilov [3]. It can be used to simplify many of their calculations. For example, let us apply it to the calculations on page 203, concerning the fundamental solution for the wave equation. Suppose M is R^3, three dimensional Euclidean space, with coordinates $x = (x_i)$, $i = 1, 2, 3$. Set:

$$f^t(x) = r - t \quad \text{with} \quad r^2 = x_i x_i.$$

Then,

$$Y^t = \frac{x_i}{r} \frac{\partial}{\partial x_i}; \quad \frac{\partial Y^t}{\partial t} = 0,$$

$$\frac{\partial f^t}{\partial t} = -1; \quad Y^t(\frac{\partial f^t}{\partial t}) = 0.$$

(6.11) takes the form:

$$\frac{\partial}{\partial t} \delta(f^t) = - \delta^1(f^t).$$

Using (6.12),

$$\frac{\partial^2}{\partial t^2} \delta(f^t) = \delta^2(f^t).$$

Set:

$$\Delta = \frac{\partial^2}{\partial x_i \partial x_i}.$$

GENERALIZED FUNCTIONS ON MANIFOLDS

Then,

$$\Delta(\delta(f^t)) = \delta^2(f^t) + \frac{2\delta^1(f^t)}{N}$$

$$= \delta^2(f^t) + \frac{2\delta^1(f^t)}{N}$$

$$\frac{\partial^2}{\partial t^2}\left(\frac{\delta(f^t)}{t}\right) = \frac{\delta^2(f^t)}{t} + \frac{2\delta^1(f^t)}{t^2} + \frac{2\delta(f^t)}{t^3}$$

Hence,

$$\left(\frac{\partial^2}{\partial t^2} - \Delta\right)\left(\frac{\delta(f^t)}{t}\right) = 2\left(\frac{1}{rt} - \frac{1}{t^2}\right)\delta^1(f^t) + \frac{2\delta(f^t)}{t^3}$$

Now,

$$\left(\frac{1}{rt} - \frac{1}{t^2}\right)(\delta^1(f^t)) = \left(\frac{1}{rt} - \frac{1}{t^2}\right)Y^t(\delta(f^t))$$

$$= Y^t\left(\left(\frac{1}{rt} - \frac{1}{t^2}\right)\delta(f^t)\right) - Y^t\left(\frac{1}{rt} - \frac{1}{t^2}\right)\delta(f^t)$$

$$= -\frac{1}{r^2 t}\delta(f^t), \quad \text{and then}$$

$$\left(\frac{\partial^2}{\partial t^2} - \Delta\right)\left(\frac{\partial(f^t)}{t}\right) = 2\left(\frac{1}{t^3} - \frac{1}{r^2 t}\right)\delta(f^t) = 0.$$

This shows, in a more direct manner than was described in [3], that

$$\frac{\delta(f^t)}{t} \qquad (6.13)$$

is (up to a constant) a fundamental solution for the Cauchy problem of the wave equation in three dimensions.

7. REMARKS ON THE BEHAVIOR OF GENERALIZED FUNCTIONS UNDER GENERAL MAPPINGS

The starting point for the remarks of this section is again a point that is implicit in the book by Gel'fand and Shilov [3], but that is not treated in detail there. (6.13) presents an elementary solution of the Cauchy problem for the wave equation in terms of generalized functions in three variables. However, on page 234 of [3] one finds another presentation of the elementary solution in terms of generalized functions of four variables (x_i, t), $i = 1, 2, 3$, namely:

$$\delta(r^2 - t^2).$$

Clearly, the connection between them should involve a formalism enabling one to "restrict" a generalized function on a manifold to a submanifold.

We might as well consider the most general problem of this sort: Suppose that M and M' are manifolds, and $\phi: M \to M'$ is a map between them. Suppose that dp and dp' are volume elements on M and M', and that H and H' are Hilbert spaces of C^∞ functions on M and M'. Suppose that:

$$\phi^*(H') \subset H,$$

and define A: $H' \to H$ as the restriction of ϕ^* to H'. Again, our basic problem can be stated as follows: Can A be extended to a mapping: $\underset{\sim}{D'} \to \underset{\sim}{D}$ of the Dirac spaces associated with H' and H? Of course, we want this extension to have certain "functorial" properties.

So far, we have been considering the case where an adjoint mapping $A^*: H \to H'$ exists. Then, as explained in Section 2, this extension of A can be defined in a very satisfactory manner. Consider the case at hand, i.e. A is defined as explained above by a map $\phi: M \to M'$. As we have seen in Section 3, if ϕ is a maximal rank, fibre mapping the adjoint mapping does indeed exist, as the "integration over the fibre mapping." In the case where ϕ is a "fibre map with singularities" the existing information is not completely adequate, but it is at least clear that a modification of the basic idea does give one a reasonably satisfactory formalism. However, in the case suggested by the Cauchy problem for partial differential equations, i.e.: where ϕ is a submanifold mapping, the basic idea breaks down completely. For example, consider the simplest case:

$$M = R^1, \text{ with a single variable } (x).$$
$$M' = R^2, \text{ with variables } (x_1, x_2).$$
$$\phi(x) = (x, 0)$$

Then, if $\Psi'(x_1, x_2)$ is an element of H',

$$A(\Psi')(x) = \Psi'(x, 0).$$

For $\Psi \in H$,

$$\langle A\Psi' | \Psi \rangle = \int \Psi'(x, 0)^* \Psi(x) dx$$

A^*, were it to exist, would satisfy:

$$\int \Psi'(x, 0)^* \Psi(x) dx = \iint \Psi'(x_1, x_2)^* A^*(\Psi)(x_1, x_2) dx_1 dx_2$$

Formally then,

$$\int \Psi'(x_1, x_2)^* A^*(\Psi)(x_1, x_2) dx_2 = \Psi'(x_1, 0) \Psi(x_1)$$

i.e. $A^*(\Psi)(x_1, x_2) = \delta(x_2) \Psi(x_1)$

This suggests a modification of the basic algebraic formalism, whereby one assumes that the adjoint operator A^* exists as a mapping: $H \to H_0'$, where H_0' is a subspace of $\underset{\sim}{D}'$ that contains H'.

However, rather than pursuing this approach in this paper, we will investigate an approach suggested by the Temple-Lighthill formulation of the theory of generalized functions [8].

Let us present the basic ideas first in an abstract form. Suppose that H' and H are (incomplete) Hilbert spaces, and that

$A: H' \to H$

is a linear transformation between them. Again define the Dirac spaces $\underset{\sim}{D}'$ and $\underset{\sim}{D}$ associated with

H' and H. Consider two sequences $(\Psi_1, \Psi_2, \ldots)$, $(\Psi_1', \Psi_2', \ldots)$ of the elements of H' and H that converge (in the weak topology) to α' and α in $\underset{\sim}{D}{}'$ and $\underset{\sim}{D}$, i.e.

a) $\langle \alpha' | \Psi' \rangle = \lim_{n \to \infty} \langle \Psi_n' | \Psi' \rangle$

for all $\Psi \in H$

b) $\langle \alpha | \Psi \rangle = \lim_{n \to \infty} \langle \Psi_n | \Psi \rangle$

for all $\Psi \in H$. (7.1)

<u>DEFINITION</u>. α and α' are A-associated if the approximating sequences (7.1) to α and α' can be chosen so that:

$$A(\Psi_n') = \Psi_n \quad \text{for all } n. \tag{7.2}$$

In this case, we use as notation (when there is little possibility of confusion):

$$A(\alpha') = \alpha$$

Let us investigate the compatibility of this definition of "$A(\alpha')$" with the earlier definition

GENERALIZED FUNCTIONS ON MANIFOLDS

in terms of an adjoint operator, i.e. we suppose that $A^*: H \to H$ exists. Define $\alpha_0 \in \underset{\sim}{D}$ by the relation

$$\langle \alpha_0 | \Psi \rangle = \langle \alpha' | A^* \Psi \rangle$$

for all $\Psi \in H$. \hfill (7.4)

Suppose $\alpha \in \underset{\sim}{D}$ satisfies (7.1)-(7.2). Now, using (7.4), (7.1a),

$$\langle \alpha_0 | \Psi \rangle = \lim_n \langle \Psi_n' | A^* \Psi \rangle$$

$$= \lim_n \langle A(\Psi_n') | \Psi \rangle$$

We conclude that, in fact, (7.1b) is satisfied with $\alpha_0 = \alpha$, and the two definitions lead to the same result in this case.

Of course, the weakness of this method of extending A to a map: $\underset{\sim}{D}' \to \underset{\sim}{D}$ is that it must be done element-by-element, and thus there is no guarantee that the extension has the functorial properties that may be required in the applications. This problem might have some hope of resolution if there were a method of constructing approximating sequences that held for a rather wide class of elements of the Dirac space. In fact, for the case that concerns us, for the applications to physics, differential geometry, or partial differential equations, H' will be a Hilbert space of C^∞ functions on a manifold, and the approximating sequences may be constructed uniformly via convolution with a sequence of functions on the manifold. (Of course, this is a technique that is extensively used in the modern theory of partial differential equations [7].)

Let us now turn to a simple geometric situation. Suppose, as proposed in the beginning of this section, that $\phi: M \to M'$ is a map between manifolds that induces the map A: $H' \to H$ between Hilbert spaces of C^∞ functions on M and M'. Suppose that in addition f is a real-valued C^∞ function on M', that N is a closed orientable submanifold on M' of codimension one on which f vanishes, and that

$$\alpha' = \delta(f), \tag{7.5}$$

i.e.
$$\langle \alpha' | \Psi \rangle = \int_N \Psi \theta, \tag{7.6}$$

GENERALIZED FUNCTIONS ON MANIFOLDS 287

where:

$$\theta = Y \lrcorner dp; \quad Y(f) = 1 \qquad (7.7)$$

Our problem: How can $A(\alpha')$ be computed, using the rules (7.1)-(7.3)? To this end, we will construct a simple-minded approximating sequence for $\delta(f)$.

Consider $p' \to f(p') = y$ as defining a map ϕ' of $M' \to R$, where y is taken as the variable on the real line, R. For $\epsilon > 0$, define:

$$M_\epsilon' = \{p' \in M': -\epsilon \leqq f(p) \leqq \epsilon\}. \qquad (7.8)$$

LEMMA 7.1. If f has no critical points on N, then:

$$<\delta(f)|\Psi'> = \lim_{\epsilon \to 0} \frac{1}{2\epsilon} \int_{M_\epsilon} \Psi' dp$$

for each $\Psi' \in H'$. $\qquad (7.9)$

Proof. It suffices to consider the case where there are functions $(f_2, \ldots, f_m)$ such that $(f, f_2, \ldots, f_m)$ forms a coordinate system for M. (The general case would then be handled by splitting up Ψ' via a partition of unity into the sum of elements of H' whose supports be in sufficiently small open subsets to which the local argument can be applied.) Set:

$$Y = \frac{\partial}{\partial f}, \quad \theta = Y \lrcorner\, dp.$$

Suppose that: $dp = j\, df \wedge df_2 \wedge \ldots \wedge df_m$. In fact, we can suppose that $j = 1$, by suitable choice of f_2. Then,

$$d(f\theta) = dp.$$

By Stokes' theorem,

$$\frac{1}{2\varepsilon} \int_{M_\varepsilon} \Psi'\, dp = \frac{1}{2\varepsilon} \left(\int_{f^{-1}(\varepsilon)} \varepsilon \Psi \theta - \int_{f^{-1}(-\varepsilon)} (-\varepsilon)\Psi\theta \right),$$

which goes to

$$\int_{f^{-1}(0)} \Psi \theta$$

as $\varepsilon \to 0$, which is equal to the left hand side of (7.9), if $N = f^{-1}(0)$. Now, for each ε, let Ψ_ε be a C^∞, real-valued function on M', which is equal to $1/2\varepsilon$ on M_ε, and equal to 0 outside of $M_{\varepsilon+\varepsilon^2}$.

By (7.9),

$$\langle \delta(f) | \Psi' \rangle = \lim_{\varepsilon \to 0} \langle \Psi_\varepsilon' | \Psi' \rangle$$

for $\Psi' \in H'$. \hfill (7.10)

With ϕ a map: $M \to M'$, $A(\Psi') = \phi^*(\Psi')$, we can now calculate $A(\delta(f))$:

$$\langle A(\delta(f)) | \Psi \rangle = \lim_{\varepsilon \to \infty} \langle \phi^*(\Psi_\varepsilon') | \Psi \rangle$$

$$= \lim_{\varepsilon \to \infty} \int_{M'} \phi^*(\Psi_\varepsilon') \Psi \, dp$$

for each $\Psi \in H$. \hfill (7.11)

Now set:

$$f_1 = \phi^*(f) \hfill (7.12)$$

$$\Psi_\varepsilon = \phi^*(\Psi_\varepsilon') \hfill (7.13)$$

Note that:

$$\Psi_\varepsilon(p) = \begin{Bmatrix} 0 & \text{if } |f_1(p)| \geq \varepsilon + \varepsilon^2 \\ \frac{1}{\varepsilon} & \text{if } |f_1(p)| \leq \varepsilon \end{Bmatrix}. \hfill (7.14)$$

Thus, we have:

THEOREM 7.2. Set:

$$N_1 = f_1^{-1}(0) = \phi^{-1}(f^{-1}(0)) = \phi^{-1}(N), \quad (7.15)$$

Suppose that:

$df_1 = \phi^*(df)$ is non-zero at each
point of N, (7.16)

Then, $A(\delta(f))$, defined by (3.11), is equal to $\delta(f_1)$, i.e. we have, somewhat symbolically:

$$\phi^*(\delta(f)) = \delta(\phi^*(f)) \qquad (7.17)$$

For the proof, one has only to remark that Ψ_ε, by (7.14), satisfies the same conditions that enabled us to prove Lemma (7.1).

8. GENERALIZED FUNCTIONS ON FIBRE SPACES

In Section 7, we have learned, in particular, how to "restrict" generalized functions on a manifold M' to generalized functions on a submanifold M. Now, we want to consider the case where M is a "moving" submanifold, dependent on certain parameters. The appropriate formulation of this idea is that of a *fibre space*. However, we will not involve ourselves with the various technical problems involved in the foundations of the theory of fibre spaces. The following set-up will be adequate for our purposes:

>M' and B are manifolds, π is a differentiable map of M' onto B, such that, for b ε B, the fibre $\pi^{-1}(b)$ is a submanifold of M'. (8.1)

We denote the fibre $\pi^{-1}(b)$ by M(b), for each b ε B. Let dp be a volume element on M', and let H be a Hilbert space of C^∞ functions on M. For b ε B, let H(b) denote the functions on M(b) obtained by restricting the functions in H to M(b). Let $\underset{\sim}{D}(b)$ denote the Dirac space of H(b), and let $\underset{\sim}{D}$ denote the *union* of all the $\underset{\sim}{D}(b)$, as b runs over B. Thus, $\underset{\sim}{D}$ may be considered as a "vector bundle" over B. Let $\Gamma(D)$ denote the space of cross-sections of this vector bundle, i.e. the space of maps: b $\to$ $\alpha(b)$ ε $\underset{\sim}{D}(b)$ which assign to each b ε B an element $\alpha(b)$ in the "fibre" $\underset{\sim}{D}(b)$ over b.

Let $\underset{\sim}{D}'$ denote the Dirac space of H'. Elements α ε $\underset{\sim}{D}'$ may be expected to define cross-sections b → α(b) of Γ($\underset{\sim}{D}$) via the restriction-to-the-fibres process described in the last section. To have an explicit method of restriction, let us suppose that:

Each fibre M(b) has a volume element differential form, denoted by $d_b q$. The restrictions of the Ψ' ε H' to each M(b) are square-integrable with respect to $d_b q$. (8.2)

There is a one-parameter family (Ψ'_ε) of elements of H' such that:

$$\langle \alpha | \Psi' \rangle = \lim_{\varepsilon \to 0} \langle \Psi'_\varepsilon | \Psi' \rangle$$

for Ψ' ε H'. (8.3)

Using (8.2), we can then make each H(b) into a Hilbert space, by integrating over M(b). Thus, if α satisfies (8.3), we can define α(b) ε $\underset{\sim}{D}$(b) as follows:

$$\langle \alpha(b) | \Psi \rangle_b = \lim_{\varepsilon \to 0} \langle \Psi_\varepsilon' | \Psi \rangle_b$$

for $\Psi \in H(b)$. \hfill (8.4)

where $\langle \ | \ \rangle_b$ denotes the inner product on $H(b)$ defined by the volume-element $d_b q$. We denote the cross-section of $\Gamma(D)$ obtained in this way by $\underset{\sim}{\alpha}$.

Of course, for certain exceptional fibres $M(b)$ the limit in (8.4) may fail to exist. One may hope to define $\alpha(b)$ for these fibres by "analytic continuation" from the b's for which the limit does exist. For example, consider the case where: $\alpha = \delta(f)$, with f a C^∞, real-valued function on M'. Let f_b denote the function on $M(b)$ obtained by restricting f to this manifold. Set:

$$N = f^{-1}(0); \quad N(b) = N \cap M(b).$$

Now, as we have shown in Section 7, $\delta(f)$ is well-defined as an element of $\underset{\sim}{D}'$, and satisfies (8.3), if:

$df \neq 0$ at each point of N.

Also, $\delta(f_b)$ is well-defined, and is equal to

$\delta(f)(b)$, providing that:

$$df_b \neq 0 \text{ at each point of } N(b). \qquad (8.5)$$

Geometrically, condition (8.5) means that N and the fibre M(b) intersect *in general position* [7]. Thus, the "exceptional" fibres for which there may be some additional complexity in the definition and properties of $\delta(f)(b)$, are those geometrically exceptional fibres which do not meet the hypersurface N(b) in general position.

Let us turn to the following question: Suppose that $\alpha \in \underset{\sim}{D}'$ satisfies a differential equation. How does one deduce the differential equations which $\underset{\sim}{\alpha} \in \Gamma(\underset{\sim}{D})$ satisfies?

The first question is, of course: What is the meaning of a "differentiable operator on $\Gamma(\underset{\sim}{D})$. This can be answered in the following general way. Let F(M') denote the ring of C^∞ functions on M'. Note that elements of $\underset{\sim}{D}'$ can be multiplied by functions in F(M'), i.e. they form an F(M')-module. Similarly, the elements of $\Gamma(\underset{\sim}{D})$ can be multiplied by elements of F(M'):

If $f \in F(M')$, $\underset{\sim}{\alpha} \in \Gamma(\underset{\sim}{D})$, then

$$(f\underset{\sim}{\alpha})(b) = f_b\underset{\sim}{\alpha}(b) \qquad (8.6)$$

where the multiplication on the right hand side of (8.6) is just the multiplication of C^∞ function on $M(b)$ (the restriction to $M(b)$ of f) by the generalized function $\underset{\sim}{\alpha}(b)$.

Let us say that zeroth order differential operator on $\Gamma(\underset{\sim}{D})$ is a complex-linear map: $\Gamma(\underset{\sim}{D}) \to \Gamma(\underset{\sim}{D})$ which results from multiplication by some element of $F(M')$, in the way just described. Proceed by induction: Suppose differential operators of order $\leq r-1$ have been defined on $\Gamma(\underset{\sim}{D})$. We can then define an r-th order differential operator as a linear map: $\Gamma(\underset{\sim}{D}) \to \Gamma(\underset{\sim}{D})$ whose commutator with *all* zeroth order differential operators is of order $\leq r-1$.

We will mainly be concerned with differential operators defined by one-parameter transformation groups. To define them, proceed as follows:

A diffeomorphism $g: M' \to M'$ is a fibre-diffeomorphism if there is a diffeomorphism $g_B: B \to B$ such that the projection map $\pi: M' \to B$ intertwines the action of g and g_B on M' and B.

Then, g maps the fibre M(b) into the fibre
$M(g_B(b))$, for each b ε B. Such a g acts on H':

$$\Psi'(p') \to \Psi(g^{-1}p')$$

for Ψ' ε H'. (8.7)

(Suppose that H' is a space of functions that is invariant under diffeomorphism, and multiplication by elements of F(M').) Since the inner product on M' is defined by integration; using a differential-form volume element, the operator (8.7) on H' defined by g admits an adjoint. The general remarks of Section 2 then provide an extension of g to an operator: $\underset{\sim}{D}' \to \underset{\sim}{D}'$. Similarly, the action g maps H(b) into H(gb) via the formula (8.7). This map also admits an adjoint, hence admits an extension to a map: $\underset{\sim}{D}(b) \to \underset{\sim}{D}(gb)$. Thus, if G denotes the group of all fibre preserving diffeomorphisms of M, G acts in a linear manner on the vector bundle $\underset{\sim}{D}$, hence also acts on the cross-sections $\Gamma(\underset{\sim}{D})$ [5]. The "Lie algebra" of G consists of the one-parameter subgroups [5]. It is readily seen that the action of those one-parameter subgroups on $\underset{\sim}{D}'$ and $\Gamma(\underset{\sim}{D})$ admit infinitesimal generators -- which are readily seen to act as first order differential operators on $\underset{\sim}{D}'$ and $\Gamma(\underset{\sim}{D})$, in the sense of the above definitions of such operators. The action of these operators obviously is intertwined by the operator $\underset{\sim}{D}' \to \Gamma(\underset{\sim}{D})$ defined by "restricting" generalized functions on M' to the fibres M_b. Thus, the same intertwining property holds for the ring of operators on $\underset{\sim}{D}'$ and $\Gamma(\underset{\sim}{D})$ generated by the zeroth order operators, and the infinitesimal generators of the one-parameter subgroups of G. This ring of operators is all that is required to deal with all the common situations encountered in mathematical physics and partial differential equations.

For example, we will deal with the fundamental solution for the Cauchy problem of the wave equation in this way. Let M' be R^4, with coordinates (x_1, x_2, x_3, t), which we will abbreviate to (x, t) with $x^2 + t^2 \neq 0$. Let $B = R$, with:

$$\pi(x, t) = t.$$

Let:

$$\square = \frac{\partial^2}{\partial t^2} - \frac{\partial}{\partial x_1^2} - \frac{\partial}{\partial x_2^2} - \frac{\partial}{\partial x_3^2}.$$

Note that, indeed, $\square$ is in the ring generated by the first order differential operators originating from one-parameter subgroups of G. ($\partial/\partial t$ and $\partial/\partial x_i$ are all infinitesimal generators of one-parameter subgroups of G.) Consider:

$$f(x, t) = x^2 - t^2 = (r+t)(r-t),$$

with $r = \sqrt{x^2}$.

$$\alpha = \delta(f).$$

Then, $\square \alpha = 0$. (P. 234 of [3]).

Identify a point $b \in B$ with a value of t. The fibre M(b) is then the space of variables (x).

$$f_b(x) = f(x, b).$$

Now, using formula (18) on p. 212 of [3],

$$\delta(f_b) = \frac{\delta(r - b)}{(r + b)} = \frac{1}{2} \frac{\delta(r - b)}{b} \qquad (8.8)$$

Now, $b \to \delta(f_b)$ is an element of $\Gamma(\underset{\sim}{D})$, i.e. a cross section of the vector bundle $\underset{\sim}{D}$. It is readily seen that the infinitesimal generator of the one group $(x, t) \to (x, t+b)$ is just given by:

GENERALIZED FUNCTIONS ON MANIFOLDS 299

$$\frac{\partial}{\partial b} \delta(f_b) = \lim_{\varepsilon \to 0} \frac{1}{\varepsilon} (\delta(f_{b+\varepsilon}) - \delta(f_\varepsilon)).$$

Similarly, the infinitesimal generator of the space translations are the operators

$$\frac{\partial}{\partial x_i} \delta(f_b)$$

defined as follows:

$$< \frac{\partial}{\partial x_i} \delta(f_b) | \Psi(x) > = - < \delta(f_b) | \frac{\partial \Psi}{\partial x_i} >.$$

Thus, the general remarks show that:

$$\frac{\partial^2}{\partial b^2} \delta(f_b) = \left(\frac{\partial^2}{\partial x_1^2} + \frac{\partial^2}{\partial x_2^2} + \frac{\partial^2}{\partial x_3^2} \right) (\delta(f_b))$$

This was deduced in Section 6 by a more explicit calculation.

Gel'fand and Shilov show [3] that, further, $\delta(f_b)$ solves the Cauchy problem for the wave equation, in the sense that:

$$\delta(f_b) \to 0 \quad \text{as} \quad b \to 0 \qquad (8.9)$$

$$\frac{\partial}{\partial b} \delta(f_b) \to c\delta(x) \quad \text{as} \quad b \to 0 \qquad (8.10)$$

where c is a constant, and $\delta(x)$ is the three-dimensional Dirac delta function. In a later work we will investigate the differential-geometric foundation of these relations. Note, for the moment, that:

$$\frac{\partial}{\partial b} \delta(f_b) = - \delta^1(f_b) 2b$$

Note also that the function $f_b(x) = x^2 - b^2$ has for, b = 0, a non-degenerate critical point, in the sense of Morse theory [9], at x = 0. This suggests that there is an intimate connection between the Morse theory of critical points of functions on manifolds and the theory of generalized functions.

BIBLIOGRAPHY

1. N. N. Bogoliubov and O. S. Parasiuk, Acta Math. **97**, 227 (1957).

2. H. Federer, Curvature Measures, Trans. Amer. Math. Soc. **93**, 418-491 (1959).

3. I. M. Gel'fand and G. E. Shilov, Generalized Functions, Vol. 1, Academic Press, New York, 1964.

4. K. Hepp, Proof of the Bogoliubov-Parasiuk Theorem on Renormalization, Comm. in Math. Phys. **2**, 301-326.

5. R. Hermann, Lie Groups for Physicists, W. A. Benjamin, New York, 1966.

6. R. Hermann, Differential Geometry and the Calculus of Variations, Academic Press, New York, 1968.

7. L. Hormander, Linear Partial Differential Equations, Springer-Verlag, 1963.

8. M. J. Lighthill, Fourier Analysis and Generalized Functions, Cambridge University Press, 1960.

9. J. Milnor, Morse Theory, Princeton University Press, 1963.

10. S. Sternberg, Lectures on Differential Geometry, Prentice Hall, 1964.

DATE DUE

GAYLORD PRINTED IN U.S.A.